D0477515

20 080 515 60

HOW LOCAL RESILIENCE CREATES SUSTAINABLE SOCIETIES

All over the world our cities, towns and communities are failing, or at the tipping point of doing so, with but a few exceptions. This is seen across the developed and developing world alike ranging from retirement age protests in France to food riots in Mozambique. The root of the problem is a toxic 'dependency culture', characterised by unjust and resource-depleting patterns of unsustainable consumption. What is needed is a dramatic new way of thinking and doing to make unstable societies resilient to future shocks – an ageing population, water scarcity, obesity and peak oil.

A taboo-shattering book, ***Local Resilience*** sets out how visionary national and local leaders can transform unsustainable societies as they attempt to recover from an age of austerity. By eliminating the culture of dependency in a socially and environmentally progressive way, the book shows how to transcend the political and social spectrum and even unify people around a common purpose. It does this by examining how leaders can make smarter interventions within complex systems to prevent the high cost of social and environmental failure arising from our current economic model. The book explores a number of contemporary themes (e.g. green economy, sustainable urban development, banking reform, equality and democratic renewal) and draws on a wealth of global case learning (e.g. Amsterdam, Brighton, Cape Town, Madison, Matara and Toyama).

Local Resilience is the sequel to the acclaimed book ***Sustainability in Austerity***, which has been endorsed by amongst others the UN and the Centre for Sustainable Urban and Regional Futures.

Philip Monaghan is an internationally recognised writer and strategist on economic development and environmental sustainability. He is Founder and CEO of Infrangilis.

Praise for this book:

'Resilience is now, in terms of local action, the fastest growing field of practice and experimentation. Hope and risk are two sides of the same coin and this book, in tune with our times, will help equip you to handle both.'

Ed Mayo, Secretary General, Co-operatives UK

'*How Local Resilience Creates Sustainable Societies* sets out in values, terms and concrete proposals how to make local communities thrive in a globalised economy. Everyone who wants to build a good society from the bottom up should read it.'

Neal Lawson, Chair of Compass, UK

'The book convincingly explains how resilience and sustainability are coming together at the local level. It demonstrates this through a wealth of stimulating innovations. This new body of local praxis is linked to an ambitious conceptual framework that can help local leaders in promoting more resilient communities and overcoming barriers to change.'

Rafael Tuts, Chief, Urban Environment and Planning Branch, UN-HABITAT

'Monaghan's follow up to *Sustainability in Austerity* captures the moment again. In these troubled economic times he demonstrates the vital direct role local government has to play in ensuring communities remain sustainable and local economies remain resilient for the long term.'

Paul O'Brian, CEO of the Association for Public Service Excellence (APSE)

'*How Local Resilience Creates Sustainable Societies* asks why orthodox approaches to sustainable urban development have failed and what can be done about this. Philip Monaghan seeks to unsettle and challenge orthodoxies and highlight the importance of transferring power to ordinary people to build resilience. He makes a practical contribution to how this can be done in an age of austerity. In doing so, he opens up a debate that will be of interest to policy makers, agencies, academics, consultants, community groups and individuals.'

Dr Mike Hodson, SURF Centre, University of Salford

HOW LOCAL RESILIENCE CREATES SUSTAINABLE SOCIETIES

Hard to Make, Hard to Break

Philip Monaghan

LONDON AND NEW YORK

For my nieces and nephews – those present and those yet to be

First published 2012
by Routledge
2 Park Square, Milton Park, Abingdon, Oxon OX14 4RN

Simultaneously published in the USA and Canada
by Routledge
711 Third Avenue, New York, NY 10017

Routledge is an imprint of the Taylor & Francis Group, an informa business

British Library Cataloguing in Publication Data
A catalogue record for this book is available from the British Library

Library of Congress Cataloging in Publication Data
Monaghan, Philip (Philip Edmund)
How local resilience creates sustainable societies : hard to make, hard to break / by Philip Monaghan.
p. cm.
Includes bibliographical references and index.
1. Community development–Environmental aspects. 2. Urban economics–Environmental aspects. 3. Sustainable development. 4. City planning–Environmental aspects. I. Title.
HN49.C6M65 2012
307.1'416–dc23
2011037355

ISBN: 978-1-84971-440-2 (hbk)
ISBN: 978-1-84971-441-9 (pbk)
ISBN: 978-0-203-12650-9 (ebk)

Typeset in Bembo
by Taylor & Francis Books

Printed and bound in Great Britain by the MPG Books Group

CONTENTS

List of illustrations viii
List of abbreviations x
List of case interviews and resilience learning xii
Acknowledgements xv
Preface xvi

Introduction: why the system is toxic: easy to make, easy to break 1

PART ONE
Ending the wrong type of dependency culture 7

1 Establishing need 9
Gaps in resiliency work to date? 9
A review of the literature 10

2 Tomorrow and today's problems: making change universally desirable 15
Dilemmas faced by global market regulators and local urban planners 15
A new paradigm for tomorrow's problems 18
Struggles to make sophisticated responses in a changing world? 21

3 A common set of values 24
No tinkering around the edges 24
Unifying beliefs 26
Significant behaviour or not? 29

PART TWO
Localism without government 33

4 **Devolving responsibility** 35
Communities on the front line or in the firing line? 35
Navigating shifts in power, rights and responsibilities 38

5 **Negotiated rights and sanctions** 42
Return to fairness through contribution 42
Area-based negotiations 44

6 **The harmonised constitution** 51
The journey from rights to responsibilities to subsidiarity 51
An enabling constitution for local leadership 56

PART THREE
Just cities 61

7 **Incentivised migration to compact cities** 63
Reaffirming the need for compactness 63
Ensuring smart density wins 67

8 **Urban development and the green economy** 72
Problem of definition 78
A decarbonised economy the North and South can believe in 80

9 **Decoupling vested interests** 86
Ending unhealthy relationships 86
Strength through diversity as well as devolution 91

PART FOUR
Transition from unstable to resilient societies: hard to make, hard to break 97

10 **Smarter and less frequent interventions** 99
Resource flows 99
Harnessing the positive power of markets and people 102
Systems thinking: from information hoarding to place-based governance 106

11 Infused resilience: a theory of change **112**
A refined interpretation of resilience 112
Embedding and maintaining empowerment 114

12 What you need to do next **118**
The right type of local leadership 118

Appendix 1 Author's biography **122**

Appendix 2 Other helpful sources of learning **124**

Bibliography 129
Index 138

ILLUSTRATIONS

Plates

1	The unsustainable society	6
2	Urban centres we all should have a stake in	32
3	From gridlock to connected cities	60
4	The resilient society	96

Figures

2.1	Rising obligations whilst frontline public services are threatened	17
2.2	Shifting governance patterns during industrialisation of cities	20
8.1	Future scenarios for green economy and urbanism?	82
10.1	Intervention matrix	103
10.2	A framework for excellence	104
10.3	Key pathways to smarter outcomes	105
10.4	Normalising action on sustainability	105
11.1	Governing sustainable urban development as a complex system	113
11.2	A theory of change to unlock the right kind of dependency	114
11.3	Infused resilience diagnostic	115

Tables

7.1	Economic power of cities	64
8.1	The interface between the green economy and urban development: so far, so good?	77
8.2	Strategic choice dilemmas for urban leaders: who to partner?	81
12.1	Partner contributions to delivering infused resilience	120

Boxes

2.1 Demands for better green regulation to boost competitiveness 19
5.1 Reforming the welfare state to provide salary insurance 43
6.1 Environmental limits legislation 57
7.1 More children killed by traffic than by diseases 66
8.1 China's five-year plan 2011–15: a switch to green growth? 73
8.2 World carbon emissions 73
8.3 Enabling the transition: expectations of local government? 75
8.4 The new 'gold rush': corporate social responsibility and urbanism 78
9.1 UK civil protests against banks spread to USA 86
9.2 Good banking gone rogue? 88
9.3 Time to break up the monopolistic energy distribution networks? 91
10.1 The city as a super organism 100
10.2 'Insuring' against cities dependencies 101
11.1 The brilliant council of the future? 115
11.2 Investing local government pension funds in decarbonisation regeneration schemes 116
12.1 World mayors sign climate change pact 119
12.2 Partner contributions to delivering infused resilience 119

ABBREVIATIONS

cCCR	Carbon Cities Climate Registry
CLT	community land trusts
CO2	carbon dioxide
CPRs	community pool resources
CRC	community resource centre
CSR	corporate social responsibility
EMS	environmental management system
ESB	environmentally significant behaviour
EU	European Union
FAO	Food and Agriculture Organisation
GDP	gross domestic product
GHG	greenhouse gas emissions
GIS	geographic information system
Ha	hectare
ICSC	International Centre for Sustainable Communities
ICT	information and communication technologies
IDB	Inter-American Development Bank
IP	intellectual property
IPCC	Intergovernmental Panel on Climate Change
IRP	International Resource Panel
MDG	millennium development goals
MFA	multi-fibre agreement
NEF	New Economics Foundation
NGO	non-governmental organisation
PLC	paralegal committee
PPP	public–private partnership
PV	photovoltaic

SO2	sulphur dioxide (SO2)
UCLG	United Cities and Local Governments
UNESCO	United Nations Educational, Scientific and Cultural Organization
UNEP	United Nations Environment Programme
WBCSD	World Business Council for Sustainable Development
WEF	World Economic Forum
WRI	World Resource Institute
WWF	World Wildlife Fund for Nature

CASE INTERVIEWS AND RESILIENCE LEARNING

Case interviews

Växjö (Sweden)
Common concerns and action on climate change

Moratuwa and Matara (Sri Lanka)
Centring women in governance

Brighton and Hove (UK)
Towards a social contract to preserve the biosphere

Cape Town (South Africa)
A constructive constitution

Tomaya (Japan)
The challenges of compact policies

Quito (Ecuador)
Ecosystems services

Madison (USA)
Community housing

Amsterdam (The Netherlands)
Food, rural-urban linkages and competitiveness

Resilience learning

Victoria (Australia)
Preventing violence before it occurs

Los Angeles (USA)
Separate lives: lessons on riots and gangs

Suffolk (UK)
Lesser government and bigger society

Havana (Cuba)
Lesser government and bigger society

Caracas (Venezuela)
Bottom-up community provision

Texas (USA)
Bottom-up community provision

Purena (Nepal)
Bottom-up community provision

Neustaudt an der Weinstrasse (Germany)
'Carrot and stick' citizenship schemes

Huyton (UK)
'Carrot and stick' citizenship schemes

San Carlos (USA)
Mobilising excluded groups in planning and campaigning

London (UK)
Mobilising excluded groups in planning and campaigning

The Basque Country (Spain)
Economic and democratic renewal through enterprise

Hebei (China)
No promotion for local officials who do not care for their family

California (USA)
Setting the nation's highest renewable power target

Totnes (UK)
Transition town

New Orleans (USA)
Saving residents money as gas prices rise

Ordos City (China)
Paying farmers to move

Lagos (Nigeria)
Changing the lives of millions through the railway

Amman (Jordan)
Refugees and the pressure of rapid urbanisation

Delhi (India)
Green does not always mean good

Pennsylvania (USA)
Green works for the economy

Asuncion (Paraguay)
Bringing informal waste collectors into the mainstream

Santa Cruz (Bolivia)
Public–public partnership for water

Lima (Peru)
Wastewater reuse for irrigation

Manchester (UK)
Local buying to save money and improve communities

Changwon (Republic of Korea)
Declaration of an environmental capital

Jätkäsarri quarter, Helsinki (Finland)
Creative use of technology

Ontario (Canada)
Creative use of technology

Medellin (Colombia)
Creative use of technology

Parades (Portugal)
Creative use of technology

Croydon (UK)
Early interventions in social care

Michigan (USA)
Place-based governance

ACKNOWLEDGEMENTS

It is important to acknowledge a number of key people who have generously contributed to this book.

First and foremost, a thank you to Eve Sadler for helpful advice and guidance on the first draft. Her painstaking review was invaluable in shaping my ideas further.

For the provision of insightful case studies sincere thanks goes to Samantha Anderson, Åsa Karlsson Björkmarker, Helen Campbell, Anton Cartwright, Pat Conaty, Thurstan Crockett, Charles Davies, Tadashi Matsumoto, Jane McRae, Patricia Miranda, Ivan Narvaez, Soraya Smaoun, Pim Vermeulen and Sumana Wijerathne.

For new thinking, their insight and for kindly sharing fantastic contacts, my appreciation goes to Raf Tuts, Neal Lawson, Ed Mayo, Mike Hodson, Halina Ward, Paul O'Brien, Gavin Hayes, David Bresch, Micel Lies, Léan Doody, Philipp Rode, Rosalie Callway, Flora Yat Ting Lim, Catherine Monaghan and Jason Hartke.

Finally, my gratitude goes to all the team at Taylor and Francis and in particular Michael P. Jones, Khanam Virjee and Charlotte Russell for backing this project, to Andrew Sadler for his great illustrations and to Chris Shaw the copyeditor.

Preface

My passion is social justice. Yet, like so many other flawed human beings in this world, pain is my 'fuel'.

I get upset when ordinary hard-working Americans lose their jobs and family homes as the arrogant bankers who caused the global financial crisis, and took taxpayer bailouts, now pay themselves record-breaking salaries by profiteering from house repossessions.

I get upset when UK war heroes who were brave enough to fight against the Nazis are too scared to switch on the heating in winter because of price hikes arising from myopic energy infrastructure planning and weak oversight of charging practices by fossil fuel suppliers.

I get upset when the Western nations shamefully look down on the East and the South as morally inferior at a time when the Arab Spring uprising shows there are still inspired revolutionaries who will fight, and ultimately are prepared to die, for the liberties of their loved ones and fellow citizens.

At the same time, I am also hugely optimistic. I am optimistic that whilst it may be awfully painful to deliver social justice for the masses the ensuing benefits will significantly outweigh the cost. Optimistic, that is, that success can be enjoyed in the long term.

This maelstrom of emotions and convictions from a career in the pursuit of social justice has turned into the fuel to write. As such, this book is part therapy, part problem-solving for me.

Core to this is the premise that transferring power to ordinary people at key points in the economic or political system will make our societies more resilient against shocks and failings, be they economic instabilities, terrorist attacks, social inequalities or climate chaos.

Doing so will require challenging orthodoxies across a kaleidoscope of disciplines – ranging from defining a universal set of values and moving beyond GDP as the primary measure of well-being, to taming the capital markets, to ensuring local decision-making is accountable to communities whilst also protecting the interests of future generations.

My intention is that this will unsettle and upset people in equal measure. It may even stir revulsion and retaliation. For some, I hope, it will inspire new thinking and action.

I look forward to the ensuing debate about how we can collectively challenge the vested interests of the few that currently undermine the will of the great many ordinary people who simply want a better life for themselves and their loved ones. Those people who, just as importantly, do not begrudge other hard-working people in their communities or the wider world a fair chance to achieve the same life goals.

If you want to join me in this movement for change join the discussion at www.infrangilis.org where you can also find the latest learning through my ongoing collaborations, conversations and developments with others.

Philip Monaghan
August 2011

INTRODUCTION

Why the system is toxic: easy to make, easy to break

Why, when it comes to sustainable urban development, has 'business as usual' failed?
Why do national and local leaders need to move away from this?
What action can municipal authorities and community champions take to create more sustainable societies?
How can they do this as the world attempts to recover from an age of austerity?
How can multilateral agencies such as the UN or OECD support the transition to a green economy through better governance arrangements?
How should the role of the private sector in new major urban infrastructure be scrutinised?

This book seeks to address these, and other, fundamental questions.

Something very important is going badly wrong. Despite decades of research, policy development and on-the-ground implementation on sustainable urban development strategies vast numbers of the world's cities, towns and communities are still failing, or at the tipping point of doing so, across the developed and developing world alike. And, at the very same time, public finances are at their worst since the Great Depression caused by debt built up from bailing out the banks. Whilst this age of austerity will not be with us forever, its impact on social policy will surely endure for a generation.

With no agreed panacea to address these challenges, the Left's calls for more state spending and the Right's vision for smaller government with a bigger society only result in further division as the former are criticised for perpetuating a 'poverty of aspiration' and the latter receive criticism for victimising the most vulnerable. Regardless of this, or political leaning, the status quo is clearly untenable and a radical 'rethink' is now inescapable.

Dramatic, bold and brave new ways of thinking are needed to make unstable societies resilient to future shocks: ageing populations, food and water scarcity, rises in obesity and diabetes rates, reaching peak oil (the point at which demand exceeds supply) or extreme weather conditions. Simultaneously, societies need to learn and evolve, notably in terms of the transition to a more self-reliant, diverse and green economy. The root of the problem is a toxic 'dependency culture'. Not in the maligned neo-conservative sense (i.e. a welfare system that encourages the poor to claim benefits rather than work), but in far-wider-ranging and classless 'addictions' such as a reliance on unjust and resource-depleting patterns of unsustainable consumption.

At its core, a radical theory of change is needed – *infused resilience* – to end dependency culture in a way that is fair and desirable (i.e. achieving greater prosperity for all and the transition to a green economy) as well as being ruthless and cost effective (i.e. less welfare spending and a reduced expense of dispersed rural–urban infrastructure such as sewers). *Hard to Make, Hard to Break* sets out a vision for how courageous national and local leaders can transform unsustainable societies as they attempt to recover from an age of austerity. The objective is to show that by eliminating the culture of dependency in a socially and environmentally progressive way, we can transcend the political and social spectrum and even unify people around a common purpose. The interventions put forward here are intended to be resource-effective by avoiding the huge cost of whole societies failing and instead nurture more viable, happier livelihoods and greener economies. This book emerged in response to this unprecedented and multifaceted challenge. It is grounded in the premise that the current economic model is fundamentally flawed in the way it discounts the high cost of failure, as evidenced by the turmoil from the latest global banking crisis.

This has been powerfully demonstrated by the pro-democracy revolutions in Egypt and Libya; public alarm at Standard and Poor downgrading the USA's credit rating for the first time in history; the retirement age protests in France; marches against public service cuts and youth mobs looting cities in England; and water conflict wars in India and food riots in Mozambique. Moreover, progress against the UN Millennium Development Goals (MDGs) is disheartening, poverty and equality gaps are increasing, climate change is out of control and rural–urban migration tensions continue to flare. Simultaneously, gigantic levels of national debt have been created by bailing out the global banking system (as witnessed to tragic effect with Greece, Ireland, UK, USA and Portugal) making public finances their worst since the Great Depression.

By illustrating how and why some societies are more sustainable than others, *Hard to Make, Hard to Break* explores the characteristics which lead to this. For instance, how resilient societies are those that through devolved power to citizens and settlements have locally agreed rights and responsibilities on the one hand (e.g. budget setting, negotiated regulation), but also a system of locally administered rewards (e.g. greater entitlements, higher social status) and sanctions (e.g. financial penalties, loss of access to amenities) to incentivise people to 'do their bit' on the other. Just as critically, these interventions are resource-effective, as they are intended to be high-level but few and far between and thus avoid the huge cost of whole communities failing and instead nurture more viable, happier livelihoods.

Hard to Make, Hard to Break has been written with multiple and international audiences in mind. Executives, mayors and councillors in local government across the world (either emerging or developed economies) will benefit from peer learning. Policy makers in central government or multilateral agencies, academics and consultants who work with the public sector on strategy, governance or delivery may find a new theory of change helpful. 'Community champions' or individuals with an interest in community participation and politics will find practical proposals to disperse power useful.

The book is divided into four parts, each building to end up with a new theory of change on resiliency:

- Part One 'Ending the wrong kind of dependency culture' (Chapters 1–3) focuses on making change universally desirable by setting out a common purpose that will unify people across a wide spectrum of social or political interests through a broad set of values, including civic duty, equality, well-being and transition to a green economy. The necessity of this type of responsible intervention is to *enable* rather than *inhibit* social mobility and prosperity for all. This is achieved through empowerment and by forging a new dependency on ecological systems (e.g. water conservation, land restoration) and social networks (e.g. families, neighbours).
- Part Two 'Localism with local government' (Chapters 4–6) then establishes the importance of devolving *responsibility* as well as *power* by setting out a framework for up-skilling local communities and public officials; alongside the development of locally negotiated regulations, rights and sanctions. Despite the name of this part, it is not actually calling for an end to local government, but rather for a new form of localism which is smarter, higher level, where less frequent intervention means more resources are diverted away from administration towards front-line delivery. This goes beyond clusters of 'co-production' (e.g. woodland management by local residents) to harmonised governance for 'total place-making' (i.e. local residents governing all priority services). At the same time, the role of the private sector is here put under scrutiny, be it in relation to major urban infrastructure projects, green patents or intellectual property (IP) harvesting.
- Part Three 'Just cities' (Chapters 7–9) refers to 'just' in both senses of the word: as a location demarcation and with regard to fairness, as cities are crucial to winning the war on poverty, inequality and climate change. Setting out why incentivised migration from dispersed rural settlements to dense and compact urban settlements needs to accompany negotiated regulation, it builds upon examples such as the average New Yorker producing 30 per cent of the emissions of the average US citizen and how women-only money-saving groups in China

are most successful in alleviating poverty and gender bias when based in the city (Barley, 2010).

- Part Four 'Transformation from unstable to resilient societies – hard to make, hard to break' (Chapters 10–12) shows how leaders can manage a sustainable transformation that is both practical and resource-effective as the world recovers from an age of austerity. This includes cautioning against single-issue policies which go against the grain of the systemic change required. For instance, the UK welfare spending reduction announced in 2011, which slashes housing benefits and may force people out of more expensive cities, could prove counter-productive in terms of the expense of reversing climate adaptation and gender equality benefits illustrated in the New York and Chinese examples above.

Hard to Make, Hard to Break draws upon primary research from case interviews with inspiring tales of resilience learning from across the world to understand the innovation behind these initiatives. Examples include new applied research with urban governance policy makers from UNEP and OECD, as well as the following primary case interviews:

- building the capacity of Sri Lankan women to lead the reconstruction of cities Moratuwa and Matara devastated by the tsunami;
- alleviating water scarcity challenges in the UK city of Brighton and Hove through a social contract to preserve the biosphere;
- developing better harmony between Cape Town's local constitution and the national constitution;
- devising compact city policies in Tomaya, Japan;
- enhancing city competitiveness through better rural–urban linkages on food in Amsterdam, the Netherlands.

This is not about reinventing the wheel. Insights from others in different contexts include stories of 'resilience learning' from wider practices to support theory throughout.

Building, shaping or embarking on a new programme of change towards enhanced local resilience does not always mean starting with a blank piece of paper. Often, it is about building on existing successes as a clear path to organisational excellence. It involves aligning activities to strategic intent and operational systems and understanding the trade-offs and related risks. All of which is explored in more detail during the course of the book.

Conclusions

- The current economic model is fundamentally flawed in the way it discounts the high cost of social and environmental failure, as witnessed by the latest global banking crisis from Ireland to the USA.

- Local leaders can play a pivotal role in breaking a toxic dependency culture to avoid the high cost of failure and create more sustainable societies.
- To succeed will require coordinated and complementary interventions amongst local council executives, mayors and elected members, national government policy makers, community champions and private contractors working in the public sector from around the world.

Aug 11.

Part One
Ending the wrong type of dependency culture

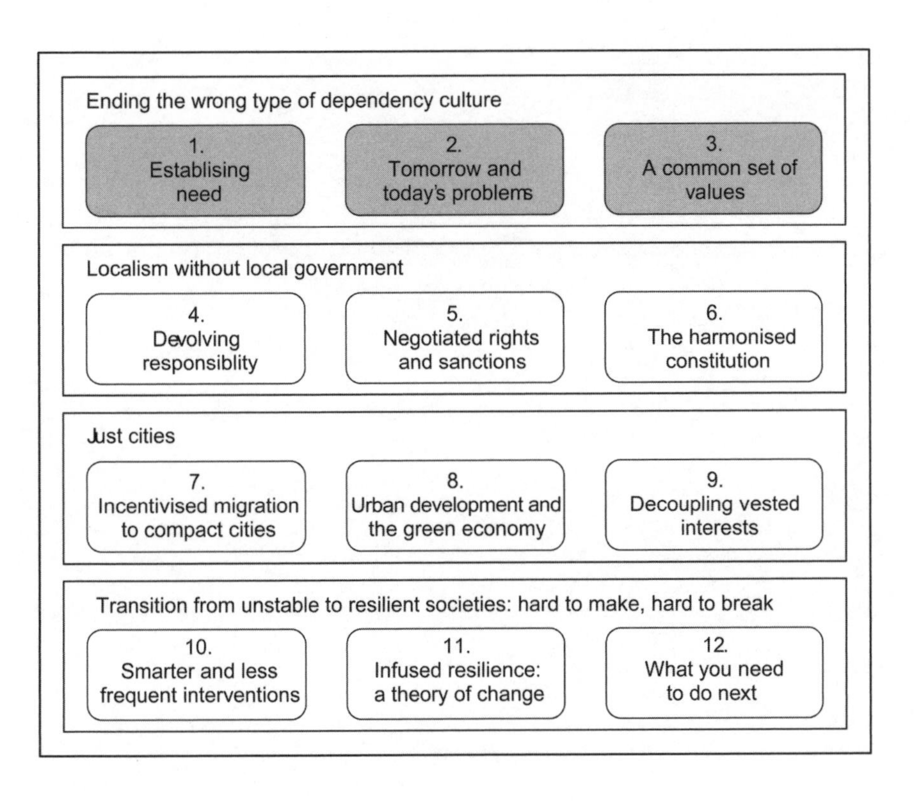

1
ESTABLISHING NEED

> Just 19 per cent of parliamentarians are women and we're more than half of humanity ... There are 19 female heads of states in 192 member states. And just 15 Fortune 5000 chief executives are women. You see we have a problem. Gender equality will only be reached if we are able to empower women.
>
> Michelle Bachelet, Executive Director, UN-Women (commenting on a gender study by the World Economic Forum that found greater productivity in countries where women achieved senior positions) (Martinson, 2011)

Gaps in resiliency work to date?

There is a vast body of relevant literature that provides helpful thinking on resilience in municipal government, sustainable communities or eco-cities and local empowerment or self-organisation. Whilst the language and terminology or points of departure vary, there is a wealth of learning about common lessons to be drawn as well as areas for further development, from which the need for this book emerged as highlighted previously.

Upon critical appraisal of these debates, a number of shared concepts or strengths can be decanted. These are as follows:

- The need to understand the underlying problems of the current economic model, how this causes unsustainable development and what needs to change – including Benello et al. (1997) *Building Sustainable Communities: Tools and Concepts for Self-Reliant Economic Change*, Khunstler (2005) *The Long Emergency: Surviving the Converging Catastrophes of the 21st Century* and Wilkinson and Pickett (2009) *The Spirit Level.*
- Using ideas or models of resilience as a pathway to more sustainable or prosperous places. These range from an ability to bounce back from shocks through to

learning and adapting, as well as the know-how to self-manage – including McInroy and Longlands (2010) *Productive Local Economies: Creating Resilient Places* and Shaw (2010) *The Rise of the Resilient Local Authority?*.
- An increasing recognition of the need for scenarios on the future of urban settlements given the vital role of cities in natural resource flows and the population's wellbeing – including Brugmann (2009) *Welcome to the Urban Revolution: How Cities are Changing the World*, Newman et al. (2009) *Resilient Cities: Responding to Peak Oil and Climate Change*, UN-Habitat (2011) *What Does the Green Economy Mean for Sustainable Urban Development?* and World Bank (2010b) *Eco² Cities: Ecological Cities as Economic Cities.*

Although these can be helpful, they also share some common weaknesses or gaps, because they display either:

- an absence of debate about a set of universal or common values which bring people or stakeholders together for the general good, or the interface between local and national constitutions to enable this;
- a lack of consistency and clarity in the use of terms such as 'resilience' and 'green economy', or how they link to the concept of 'sustainability', which results in narrow or silo interpretations that make the case for universal change difficult to argue;
- a limited analysis of the need for a different type of governance to set standards and monitor implementation, ranging from the poor financial health of so-called 'greenest cities' (i.e. predications that some of these cities will 'go bust': Moya, 2010) through to the accountability of major urban infrastructure schemes (e.g. rapid transport and energy distribution networks) that involve new and complex forms of public–private partnerships.

Each of these will be discussed in subsequent chapters and will culminate in the theory of change in Chapter 11. To reach a fuller understanding of the relevance and gaps to the debate, some of the key texts are discussed in more detail below.

A review of the literature

Shaw (2010) provides a fascinating short history of the term 'resilience' and its changing use in local government. Citing Holling (1973), Shaw explains how the term was initially used in an ecological context and defined as the 'measure of the persistence of systems and of their ability to absorb change and disturbance and still maintain the same relationships between populations or state variables'. The concept was subsequently taken up with regard to civic emergencies such as disruption to utility suppliers or a terrorist attack. More recently, it has developed to focus on how individuals and communities cope with external stresses and disturbances caused by environmental, social or economic change such as climate change or financial recession, notably in North America, Australia and New Zealand. Referring to Adger (2010) Shaw argues that resilience can be viewed as involving three elements: the ability to absorb perturbations and still retain a similar function; the ability of self-organisation and the capacity to learn; and the ability to change and to adapt. Shaw concludes that the maturation of the use of the concept of resilience amongst local authorities should be welcomed, but cautions that further research is needed to ensure consistency in its understanding and its application. Despite these strengths, Shaw does not go on to discuss the underlying system's failure which causes these problems and which needs to be addressed.

McInroy and Longlands (2010) adopt a different approach to Shaw. Here an emphasis on resilience is very much on local economic productivity and sense of place. Coined by them as the 'boing factor' (in terms of an ability to bounce back from adversity), resilience 'differs from sustainability in that it focuses on the proactive capabilities of a system to not simply exist but instead survive and flourish' (ibid.: 13). Interestingly, McInroy and Longlands cleverly put forward a model with measures to assist practitioners, which is very helpful although somewhat limited given that only one of the eight measures relates to the environment and that it focuses solely on climate change, therefore discounting the high cost of the wider mishandling of natural resources such as watercourses or agricultural land. Also, similar to Shaw, the authors do not discuss in any depth the flawed wider systems that lead to an absence of resilience, beyond a reference to local government mismanagement.

In contrast, Benello et al. (1997), Wilkinson and Pickett (2009), Kunstler (2005) and Newman et al. (2009) do attempt to conceptualise the cause of failure, but with quite different conclusions. For Benello et al. the absence of 'bottom up' economic leadership is the critical failure of unsustainable communities, whereas for Kunstler, and also for Newman et al., the fossil fuel dependency of cities and the world in general is the root cause of the problem.

Benello et al. (1997) – a seminal book based on the Schumacher Society Seminars of 1982 on community economic transformation – intend to provide essential concepts and institutions for building sustainable communities. They argue that the two basic causes of economic breakdown from 'top down' transformation (concentrated wealth, alienation between workers/consumers/residents, lack of self-management from big government, lack of finance for small or local businesses, trade wars and unemployment and welfare payments) are the rules used for owning land and corporations and the government-created monopoly money systems supported by central banking. Consequently, three major solutions identified to avoid this high cost of failure are: (i) community land trusts and other forms of community ownership of natural resources (e.g. housing, forests), (ii) worker-managed enterprises, and (iii) community currency or banking. In support of their argument Benello et al. cite the impressive example of the Woodland Community Land Trust in Tennessee, USA (a vehicle for residents to regain control of lands exploited by corporate owners and then abandoned). In addition they highlight the Seikatsu Club in Tokyo, Japan, a consumer cooperative with an emphasis on ecological sustainability and consumer control of the marketplace (e.g. food, clothes etc).

In putting forward these high-level concepts and tools for self-reliant economic change, perhaps it would be helpful for Benello et al. to produce a supporting narrative, in terms of the common purpose or values that citizens or communities would sign up to in order to deliver against this. The authors' conclusions are clearly influenced by Schumacher's 'small is beautiful' doctrine and so may also be limited by their relevance to today's urban reality as they do not offer ways to 'scale-up' to city-level interventions. This is particularly relevant when considering that half the world's population lives in cities (Brugmann, 2009). Neither did the authors extend conclusions to major urban infrastructure upon which locally owned natural assets may be dependent (e.g. water or energy distribution networks) and which in isolation or combination undermine the authors' core proposition of 'self-reliance'.

Wilkinson and Pickett's (2009) groundbreaking analysis of the impact of how inequalities hamper our shared prosperity correlates with much of Benello et al.'s thinking on the causes of unsustainable communities. Citing the links between income inequality and the financial crashes of 1929 and 2008, they argue that greater equality is in everyone's interest as it reduces the need for big government. For example, there are more police, prisons, health and social services in the USA and UK in comparison to Japan and Norway – and so they are more expensive to maintain. (The cost–benefits of avoiding these failures are explored in more detail in Chapter 2).

Newman et al. (2009), by comparison, recognise the vital role cities play in human resilience, but examine it through a response to peak oil and climate change. Kunstler (2005) also focuses on the emergency of fossil dependency, although with only a passing reference to the city. Newman et al. argue that resilience is destroyed by fear and built upon hope. Citing the examples of New York after 9/11 and London after the 7/7 bombings, the authors argue that fear stopped people from congregating on the streets or using the underground but that both steeled themselves to carry on, for instance in the case of the latter with billboard signs declaring '7 million Londoners, 1 London' organised by the city government in the immediate aftermath. Extending the reasoning, Newman et al. went on to conclude that finding hope in the steps that must be taken to create resilient cites in the face of peak oil and climate change is important. The context to this is that cities have grown rapidly in an era of cheap oil since the 1960s and 1970s and now together consume 75 per cent of the world's energy and emit 80 per cent of the world's greenhouse gas emissions; at a time when 'peak oil' – the maximum rate of production of oil, recognising that this is a finite natural resource, subject to depletion (Newman et al., 2009) – will be reached. In the UK and USA the estimated date to reach peak oil is somewhere between now and 2020. And that some cities are less well adapted and vulnerable to change, for instance, Atlanta, USA needs 782 gallons of gasoline per person each year for its urban system to work, but in Barcelona it is just 64 gallons (Newman et al., 2009). Looking at scenarios for the future of cities, Newman et al. call for the resilient city to be shaped by a 'sixth wave' of industrialism – the beginning of an era of renewable and distributed technologies for heat, power and transit which are facilitated by higher-density, 'walkable', mixed-use developments that are integrated through a 'smart grid'. This would be good news for everyone in terms of economic security, better health and an improved environment. (Scenarios for the future of cities are something we return to in Chapter 2.)

Similarly, Kunstler (2005) argues that our dependency on fossil fuel is one of a number of converging catastrophes of the twenty-first century in addition to peak oil and over population. Although, in contrast to Newman et al.'s hopeful message, Kunstler argues that we are sleepwalking to the future. Kunstler's assertion is that:

> We have walked out of our burning house and we are now headed off the edge of the cliff. Beyond that cliff is an abyss of economic and political disorder of a scale that no one has ever seen before.
>
> *(Kunstler 2005: 1)*

One conclusion he draws from this is that 'nuclear power may be all that stands between what we identify as civilisation and its alternatives' (ibid: 146) on the basis that other renewable options are not currently viable.

A downside to both Newman et al. and Kunstler is an absence of debate about the governance of these major new urban infrastructures, be they smart grids or other technologies. (This is an issue which is analysed in more depth in Chapter 7.)

Similar to Newman et al., Brugmann (2009) places the role of city planning at the centre of human progress. However, his point of departure is social and economic development to address these problems as opposed to Newman et al.'s primary focus on environmental sustainability. 'Cities are changing *everything* ... for better or worse' states Brugmann (2009: ix), noting that half the world's population now live in cities, equivalent to 3.5 billion people, and will be joined by a further 2 billion people over the next 25 years. He argues that success in a world increasingly organised around cities requires a new practice of design, governance and management called 'urban strategy'. Citing the examples of Barcelona (Spain), Chicago (USA) and Curitiba (Brazil), Brugmann concludes that 'strategic cities' like these need to master the 3 faculties of: (i) a stable governing alliance with the power and resources to align interests in the pursuit of a common goal; (ii) an explicit and detailed body of practices to translate this vision into practical forms of building and design; and (iii) a set of dedicated institutions with the technical talent to implement these solutions.

Whilst a formidable narrative, the comparative weaknesses of Brugmann are: (i) the lack of detail around what a common set of values may be to shape any common advantage; (ii) only a passing reference to the environmental limits of this economic and social development; and (iii) the absence of any debate about the questionable financial standing of these 'strategic cities', which suggests a disconnect with the failures of the underlying economic model (noting again that some commentators say 100 US cities face financial ruin: Moya, 2010). Again, these are topics we will return to in subsequent chapters of the book.

The World Bank (2010a, 2010b) and UN-Habitat (2011) both take the importance of cities a step further (albeit in quite distinct ways), arguing that we need to better understand the interface between city planning and material or resource flows; that is, to ensure there is an integrated approach that allows for better resource efficiency, life-cycle analysis in investment decision-making and greater social equity. One key difference between the two reports is that UN-Habitat strongly emphasise the interface between sustainable urban development and the 'green economy', and in particular the need to more clearly define the term (e.g. is it about uncoupling carbon from growth or prosperity?) as well as the good governance of major new urban infrastructure under this banner involving public–private partnerships (PPPs) which will shape the future of entire cities for generations. Which, interestingly, is a theme that is absent from important work on the green economy by the lead agency UNEP (2011b) thus far. The green economy is a concept we will return to in Chapter 7.

As well as this literature, other sources of critical insight, such as applied research amongst trade associations and practitioner networks such as APSE or ICLEI–Local Governments for Sustainability, is also recommended. More details are contained in Appendix 2.

Conclusions

- There is a wealth of learning in resiliency theory and practice. In particular there is an impressive body of work on the underlying problems with the current economic system, with an increasing recognition of the importance of urban planning in creating less fragile societies.
- Gaps still remain, however. Notable deficiencies include an absence of any universal values which bring people together for the common good, a lack of clarity or consistency on what is meant by the term 'resiliency' and a limited analysis of the governance arrangements needed to deliver this change including oversight of the role of the private sector in urban development.

2

TOMORROW AND TODAY'S PROBLEMS

Making change universally desirable

> Name me a city that was not built on industrialisation. The green economy can be a new wave of sustainable urban development. ... But we need to be able to sell it to people who just want to own their first toilet. And for the 'common good' to prevail cheats must not prosper.
>
> Joan Clos, Executive Director, UN-Habitat, 2011

Dilemmas faced by global market regulators and local urban planners

We need to fundamentally change our current economic model. It is a model which unfairly discounts the high cost of social failure yet handsomely rewards the mismanagement of finite natural resources. This is bad news for everyone, because, in doing so, this model makes our societies less resilient to life's shocks and surprises.

In 2000, all member states of the UN signed up to the Millennium Development Goals (MDGs), committing to targets by 2015 on issues such as poverty and hunger, education, gender equality, child mortality, maternal health, HIV/Aids and other diseases, climate change and access to technology. According to Burke (2010), whilst in the last 10 years there have been some achievements, the MDG report card is mixed. Failure to advance the interests of women and to live within environmental limits is particularly bad.

Turning first to the status of women, feminists such as Banyard (2010) shine light on the 'equality illusion' (the misguided illusion that gender equality has now been accomplished) and show how women and men still remain shockingly unequal in the developed and developing world alike. Banyard brings forward statistics showing that women own just 1 per cent of the world's land and property, that two-thirds of the world's illiterates are female and that one in every four women in the UK will experience violence at the hands of a current or former partner.

With regard to the failure to meet environmental limits, in an alarming scenario Heinburg (2007) describes that we are at 'peak everything', where the twentieth century has witnessed unprecedented growth in population, in energy consumption and also in food production. This has ushered in an era of decline in a number of crucial parameters: fossil fuels, grain harvests, climate stability, fresh water and minerals and ores. To adapt to this profoundly different world, we must begin now to make radical and practical changes to live within new limits dictated by nature.

This point is further brought into focus when considering the view of the World Wide Fund for Nature (WWF) which highlights that humans use resources equivalent to those of 1.5 planets (Jowit, 2010); and subsequently according to the IPCC developed countries' carbon cuts since 1990 have been cancelled out 3 times over by increases in imported goods from developing countries such as China (Clark, 2011). Moreover, says the UN, climate change could derail years of progress in improving the lives of the world's poor (Elliott and Tran, 2010). Other research by Beamant (2011) citing the work of Unicef, Plan International and Save the Children also highlights the systemic nature of this problem, for instance how worsening climate change is increasing the number of disasters such as droughts and floods which are hitting children in poor countries the hardest.

Brown (2011) concurs with and builds on this line of argument. He depicts 'a world on the edge', faced with a new tide of environmental refugees (e.g. 300,000 evacuees from New Orleans' Hurricane Katrina in 2005 do not plan to return) and the inability for us to adequately feed the planet's 8 billion people (e.g. he cites the ban on grain exports by Russia – the world's third biggest wheat exporter – to rein in soaring domestic food prices). This is a world, according to Brown, which can only be saved through the advancement of an affordable 'Plan B'. The cost of saving the planet through basic social goals (universal primary education; the eradication of adult illiteracy; adequately supporting women, infants and pre-school children; ensuring maternal health and family planning; and universal basic health care) and through environmental restoration goals (protecting topsoil or cropland; planting adequate numbers of trees; restoring rangelands; restoring fisheries; stabilising water tables; and protecting biological diversity) would be equivalent to only 12 per cent (US$185 billion) of the world's military budget (US$1,522 billion), Brown argues.

Yet, as worthy and important as this 'depletionist' thinking is, this is a limited solution for three reasons. Firstly, the focus is on *mitigating* as opposed to *changing* the current failed economic model, with particular oversight to global financial trading. Secondly, whilst macro interventions are necessary this is not sufficient and locally led problem solving, notably through municipal authorities and community groups, is also necessary. Finally, and building on the first two points, our leaders need to come to terms with new governance arrangements required to ensure accountability of the increasing role of the private sector in leading major urban infrastructure projects, which are intended to build more sustainable societies.

The full extent of the failed economic model is laid bare by the current global financial crisis and the roles of the banks in it. In recognition of some of the systemic failings, the European Union in 2011 announced an investigation of 16 of the world's leading banks (including HSBC, Royal Bank of Scotland and Barclays) over suspicion of collusion and abuse of their positions in providing the financial products blamed by some for exacerbating the eurozone debt crisis (Goodley, 2011). Despite this, many of the banks that took government handouts and were partly responsible for the financial meltdown of 2008/9 have subsequently announced record profits with staff remuneration on the increase by 2010/11 (Rushe, 2011). Traders are side-stepping the contentious bonus payouts, and further fuelling public outrage, by more than making up for their loss through a permanent rise in basic salaries (Bowers, 2011).

So when considering that a common consequence of national austerity measures has been to cut local spending on front-line public services – between 10 per cent (US) and 28 per cent (UK) from 2011 onwards – this is akin to throwing a brick through a neighbour's window and then asking them to pay for the cost of your brick! Worse still, it appears that some of the same speculators – including Barclays, Goldman Sachs and Morgan Stanley – that were to blame for the sub-prime mortgage crisis are now at centre of another, this time concerning food commodity price rises (Lawrence, 2011).

Yet, perversely, threats to front-line public services hitting when climate change is out of control and higher costs of living are forcing more families into poverty is nonsensical. One interpretation of this 'perfect storm' (Monaghan, 2010) is depicted in Figure 2.1.

To contextualise this, Narey (2011) shows that failure to tackle child poverty costs the USA 4 per cent of its total gross domestic product (GDP) through lost tax revenues and in additional spending on health and criminal justice. In addition, Narey

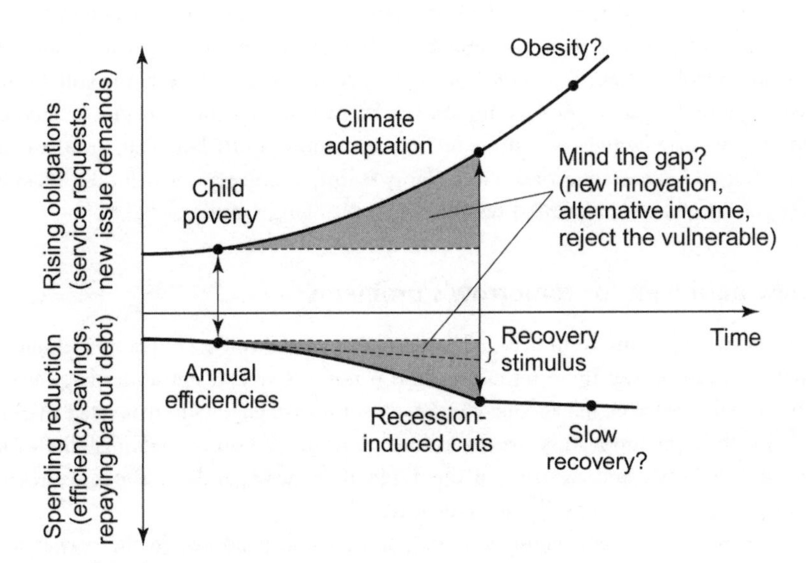

FIGURE 2.1 Rising obligations whilst frontline public services are threatened (© Monaghan, 2012. Adapted from Monaghan, 2010)

highlights that the UK government could avoid US$41.2 billion-worth of annual spending if poorer children were given a better start in life.

In terms of locally led problem solving, urban areas are a key part of the answer. Cities account for both the majority of the world population and global emissions. By 2040, two in every three people on the planet will live in urban areas. Yet there is little time to make major changes to ageing city infrastructure that has often been around for centuries given the great migration is already underway (Kaplan and Simpson, 2011). Coupled with the fact that municipal budgets have been slashed to pay for the bank bailout (to the extent that it is predicted that over 100 American cities face financial ruin as a result of the global financial crisis; Moya, 2010), and that, ironically, many are turning to the private sector to help plug the gap. This in turn may create new challenges as well as presenting new opportunities.

Over the past few years a plethora of business-inspired 'smart' cities initiatives have emerged, increasingly under the banner of 'corporate social responsibility' (CSR) or the 'green economy'. These range from the Home Depot Sustainable Cities Initiative (Home Depot) to the WBCSD Urban Infrastructure Initiative (World Business Council for Sustainable Development). The rising importance of these companies in local government circles is highlighted to startling effect in a report listing the top 50 most influential voices in UK local government, which ranked a senior executive at telecoms giant BT at number 28 (LGC, 2011). However, it is unclear how such private-sector-led or -dominated schemes are accountable to municipal leaders or the citizens they represent. After all, being pro-business and anti-weak governance are not mutually exclusive. (This interface between urbanisation and the green economy is explored in detail in Chapter 7.)

So, having considered inequality of the economic model, the rise of the importance of the city and a potential absence of regulation of private sector involvement in urbanisation, it becomes evident that change is desirable, whereby vested interests at the market or municipal level are taken to task, the high cost of social failure is recognised and power wrestled back to local communities. From this two things begin to emerge. First, a new paradigm is occurring on the interconnections between local leaders, the market and citizens. Second, more clarity is forthcoming on how local leaders are struggling to make sophisticated responses in a changing world.

A new paradigm for tomorrow's problems

As we have now come to understand, cities play a crucial role in human development and the stewardship of finite resources. This is not a new concept, indeed six 'waves of industrialism' have shaped our cities for centuries argues Newman et al. (2009) whereby through innovation we have moved from traditional walking cities with some use of horses and carriages in the 1700s (first wave) and are about to enter a new era of electric transit systems (sixth wave).

Yet, at the same time, whilst insightful, this analysis is limited in the way it does not appreciate the multifaceted nature of the problems local leaders face. For instance, it omits an assessment of how the ever increasing role of private money in urbanisation is impacting on the public's ability to scrutinise new forms of governance that arise from this. For instance, one study estimates that about 50 public–private partnerships (PPPs) were operating in the 1980s and by 2006 this had mushroomed eightfold to at least 400 (McKinsey and Co., 2009). And again, at the same time, public trust in how big companies are operating in the best interests of society is low (Zadek and Merme, 2003). But is this public distrust always being matched by public scrutiny over governance, or is there a time lag and there sometimes has to be a major scandal first which acts as a wakeup call for reform? This could indeed be true, as depicted in Figure 2.2, which adapts the 'waves of industrialism' by Newman et al. (2009) and Hargroves and Smith (2005).

The figure allows us to observe change over time from a governance perspective during these waves of industrialisation. Over the centuries there have been tremendous examples of how private sector innovations have transformed society for the common good ranging from steam power (second wave), electricity (third wave) and aviation (fourth wave) to biotechnology (fifth wave) in terms of widening access to affordable travel, power and medicine respectively. Big interventions by responsible companies continue to this day too, for instance with a number of progressive leaders in the business community over the past half decade calling for responsible law-making to enable a greener economy. One example is a study (see Box 2.1) by the Aldersgate Group – an alliance of leaders from business, politics and society whose members include BT, Veolia and Pepsi – which is demanding better green regulation.

Box 2.1 Demands for better green regulation to boost competitiveness

According to the Aldersgate Group (2006), increasing globalisation of the world's economy has raised concerns about the ability of UK industry to compete in global markets, which has led to illogical demands by industry lobbyists that UK and EU environmental regulation should be abandoned to ensure a level playing field with rising competitors in the East.

On the contrary, the Aldersgate Group argue that high environmental standards can contribute to economic competitiveness and stimulate wealth creation, indeed that they are actually mutually reinforcing. This is based on assumptions that the UK's long-term economic success depends on maintaining a healthy environment and the sustainable use of limited natural resources. For instance, they estimate that UK industry could save up to US$6.2 billion in annual operating costs by investing in best practice manufacturing techniques for waste minimisation and through energy efficiency. Higher environmental standards would also reduce the cost of damage to economically important ecosystems such as forestry and building materials such as stone and rubber (e.g. the impact of volatile organic compound emissions – such as from automobile fumes – is estimated to cost up to US$576 million per year).

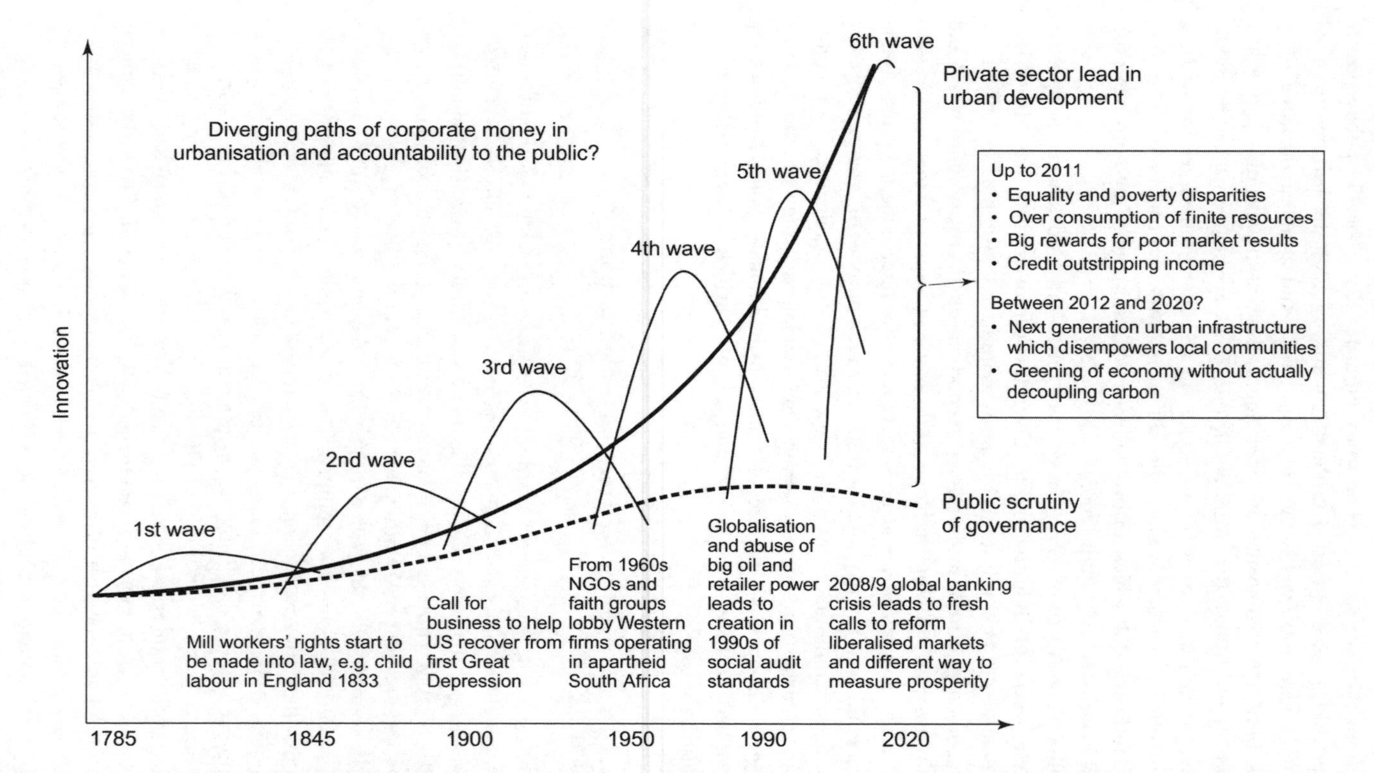

FIGURE 2.2 Shifting governance patterns during industrialisation of cities (© Monaghan, 2012. Credits: Newman et al., 2009, Hargroves and Smith, 2005)

But over the centuries there have also been incidences of corporate malpractice, which has led to the process of public scrutiny being revisited to ensure it remains fit for purpose. This ranges from concerns over mill workers' rights in the 1800s to more recent accusations about the abuse of big oil developing countries during the 1990s and present day worries about the unabated power of the capital markets in the wake of the subprime mortgage crisis.

It seems that we may be about to enter new troubled waters, with a further widening of the accountability gap in how the markets are involved with major new urban infrastructure projects (e.g. asset ownership, supply competition, profit sharing, contract flexibility, equality of access, decoupling carbon from growth, etc). And this is when there is a great migration to cities, where most people will live and most resources will be consumed. Surely this raises big questions over how inadequate governance may undermine the resiliency of many of these urban centres, indeed most of the inhabited planet? In short, the answer is yes (these are concepts which we explore in more detail in Chapters 8 and 9).

Struggles to make sophisticated responses in a changing world?

Despite the big dilemmas faced by local leaders in dealing with such multifaceted problems, there are a number of forward-thinking local leaders who are attempting to reverse some of these trends, prevent the high cost of social failure and improve their resiliency therein. But this is proving an enormous challenge.

One example is the inspirational work of the state government of Victoria, Australia, in tackling violence against women. This again includes looking at the role of business, but this time from a responsible employer perspective. The context to this smart intervention is that UN Women believe there is no better tool for development than the empowerment of women. According to Martinson (2011) this line of argument is further supported by financial bottom line and human rights arguments (this has been evidenced by a study from the World Economic Forum (WEF) amongst others; see Martinson, 2011). As the Victorian case shows, there is a huge financial benefit of ending violence against women (e.g. in the USA violence against women costs US$5.8 billion a year in terms of medical costs, loss of productivity and childcare). In Australia they have estimated that it costs US$14.4 billion, more than the US$10.6 billion they spent stimulating the economy (Martinson, 2011).

Preventing violence before it occurs in Victoria, Australia

Victoria is the second most populous state in Australia, with over 5.5 million residents. The state includes 32 cities and 39 shires, with 70 per cent of all citizens residing in Melbourne, the capital city. Victoria is also the second largest economy in Australia, accounting for a quarter of the nation's GDP. Finance, insurance and property services form the state's largest income producing sectors, while the community, social and personal service sectors are the state's biggest employers. Despite the shift towards service industries, the declining manufacturing sector remains Victoria's single largest employer and income producer. Consequently, Victoria has the highest unemployment rate in the country.

Victoria has a high rate of violence perpetrated against women and a vast body of evidence demonstrating the serious social, economic and health consequences of violence for individuals, families and communities more broadly. According to the Victoria State Government, violence against women was prevalent, was the most significant risk factor for the health of women aged 15–45 years and cost the Australian community billions of dollars (VicHealth, 2007).

The prevention of violence against women was identified as a priority for Victoria in 2004 (and the resulting state plan was the first of its kind in Australia).

The evidence presented in the plan suggests that action to prevent violence against women is best guided by three interrelated themes:

- promoting equal and respectful relationships between men and women;
- promoting non-violent social norms and reducing the effects of prior exposure to violence (especially on children);
- improving access to resources and systems of support.

To address these themes, primary prevention efforts are most likely when a coordinated range of mutually reinforcing strategies is targeted across these levels of influence. These include:

- programmes to reduce individuals' risk of perpetuating or being subject to violence;
- broad-scale social marketing campaigns to shift relevant attitudes and community norms;
- interventions to strengthen the capacity of communities, organisations and corporate action;
- reform of relevant policies and legislation.

Engaging with local and regional communities was key to success. Engagement across councils in the state included:

- Maribyrong City Council – undertaking professional development with council staff and other stakeholders to integrate violence prevention into new and existing policies, programmes and activities in a range of community settings.
- Wyndham City Council – the engagement of pupils from local schools in the production of artwork promoting messages related to preventing violence against women.
- Moreland City Council – engaging local businesses in education and training activities to raise awareness and understanding of family violence and support changes in workplace practices and policies to prevent violence against women.

For further information visit http://www.vic.gov.au (accessed 24 April 2011).

Whilst evidence of the success of this intervention is yet to emerge, it remains an insightful example of how local leaders are striving but struggling to make their communities more resilient by preventing complex problems occurring in the first place through proactive and targeted measures. (How to deal with complex systems in the smartest way is a key area of investigation in Chapters 10 and 11.)

This focus on attitudes and community norms raises a broader question, though, about how to bring people with different values together for the common good, and especially so in a changing world; that is, how a universal set of values could be developed and accepted. This is the focus for the next chapter.

Conclusions

- National and local leaders are faced with a 'perfect storm' of rising sustainability problems and less money in a recession but also public scepticism about the need to take action on certain issues like climate change.
- The rise of the importance of cities means we need to rethink how best to deal with these multifaceted problems.
- A new paradigm is emerging whereby the private sector is increasingly influencing urban policy, which, whilst it may be helpful, must also be accountable.

3

A COMMON SET OF VALUES

> No more is Middle East a synonym for conflict. The people's will has revived a sense of common destiny.
>
> Ahmet Davutoglu, Turkish Minister for Foreign Affairs, commenting on the 'Arab Spring' (Davutoglu, 2011)

No tinkering around the edges

Eighteenth-century Paris was notoriously one of the filthiest cities in Europe. The general population were enslaved in disease-ridden industries and suffered grotesque poverty in servitude to the rich elite. These great injustices were part of the contributory causes which drove Parisians to rise up and lead a bloody revolution against their oppressors which would transform the capital city and give birth to a new French republic (BBC, 2011a).

Two hundred years later and other democratic revolutions have resulted in huge change, this time affecting a whole region with protests and demonstrations leading to revolutions across Bahrain, Egypt, Libya, Syria, Tunisia and other parts of the Middle East. Here again ordinary people, just like those Parisians over 200 years ago, are willing to sacrifice their lives for the hope of something better for themselves, their families and, ultimately, their countries (a revolution which also saw Al-Qaida's influence slip from marginal to almost irrelevant in light of this 'Arab Spring', with many reporting that during this bypassing of the forces of militant Islam social networking like WikiLeaks, Twitter and Facebook had more effect: Black, 2011; Walker, 2011).

It is such brave and courageous waves of popular change both throughout history and in recent memory that make some of the fixes posed to address contemporary injustices appear worryingly meek by comparison. For instance, in the wake of the savage recession caused by the 2008/09 global debt crisis in Europe and the USA, some church leaders in the UK proposed that bankers and politicians who had caused this devastation should be legally required to work with the poor and needy to remind them of the purpose of their power and wealth (Butt, 2011). However, whilst this would undoubtedly prove amusing television viewing to watch the most powerful people in the country clean floors of a residential care home, surely this flimsy philanthropy is just tinkering around the edges? Especially when compared to the truly revolutionary spirit that caused real change in the Arab Spring?

Perhaps not surprising, though, given the steady erosion of hard-won civil liberties related to the right to peacefully protest against ruling government in the UK, USA and elsewhere, as alleged by a number of respected social commentators and historians such as Grayling (2007). This is most notable in the wake of the 11 September 2001 atrocities, in the name of defence against terrorism, whereby there has been a swage of legislation that permits an increase in government secrecy, greater breaches of personal privacy and higher incidences of imprisonment without trial (Grayling, 2007). Activities, say campaigners, which of course make it more and more difficult for ordinary people like them to protest against government policy be it austerity spending cuts in Greece, changes to pension rights in France or student fee charges in the UK as we saw between 2009 and 2011.

Pro-government voices counter that those who argue against such measures are either naive, as this temporary erosion is a price worth paying for security against terrorism, or that protestors are unpatriotic during a time of modern war. Civil liberties campaigners point out that we should direct our criticism at what democratic countries are doing to their own people as this damage is much greater than any threat of terrorism as it harms millions, not hundreds or thousands, and that, furthermore, it means the terrorists' campaign of fear is being won as it is forcing us to harm ourselves.

And so, not only is the Middle East no longer a watchword for terrorism or an absence of moral courage; instead the opposite may now be true with brave protestors fighting for their liberty and the will of the ordinary people of the Arab world having triumphed. But, more than this, the revolutions ultimately should force deep self-reflection in the West, which is now challenged to value this amazing push for liberty, equality and representation over and above any limited and selfish notion of regional security (Milne, 2011).

Indeed, some commentators go further and also argue that the West should stop imposing its 'one-size fits all' theory of democracy on the East and the South (Heine, 2011) which demonises elected leaders (e.g. from Bolivia and Ecuador), and sets the political context against which international development funds are disbursed and trade rules negotiated. More than this, Heine (2011) and others demand the West cease to believe that democracy has exclusive roots and forms from their own countries and instead for the West to open up to new ways of people organising and taking charge of their own destiny.

So, given the similar injustices that led to revolutions in France, the Middle East and elsewhere, there could indeed be common values between ordinary people who all want liberty, equality and representation. That is, instead of focusing on what makes us different, are there unifying beliefs that can bring us together?

Unifying beliefs

To begin to understand if unifying beliefs is indeed possible requires a journey of learning, starting with individual 'happiness', before moving on to common 'fairness' and ultimately to a debate about power 'devolution'.

Attempts to move beyond the traditional measure of wealth, GDP, to instead measure well-being or happiness are not new (something which is explored in subsequent paragraphs). The global debt crisis and economic depression of 2008/09 appears to have just reinvigorated this interest. Put simply, the main shortfall of GDP is that not only does it not use all applicable financial measures, but that it may actually be measuring the wrong things anyway. Take for instance an oil spill, which creates the need for clean-up services which increase GDP whilst the local environment is damaged and people's quality of life decreases (Stratton, 2010). Whilst no one is disputing the fact that wealth remains preferable to poverty, money may not be the only thing that makes us happy. In fact, Ramesh (2010a) shows that Europe's emotional well-being has dropped as incomes have risen.

Consequently, the governments of Canada, France and the UK all now have some measures of well-being to sit alongside GDP, and the OECD has a 'Better Life Index' to allow the public to judge overall standards of living in its 34 member countries (OECD, 2011). These indexes include measures ranging from how much time people spend in traffic jams, to the ratio of working hours with leisure time, to the fear of crime, to whether men and women are treated fairly in the workplace and to gender roles in the home. Whilst the measures may vary from country to country, it is still thought-provoking to ponder the results of a happiness index for nations in Europe by Gallup (Elliott, 2010), which gave countries a percentage rating (the higher the rating the happier they were said to be). In this study data, Gallup reported that the Danes (82 per cent) are the happiest people, whilst the Germans are least happy (43 per cent).

Yet, what appears to be a recurring feature of most unhappy societies, according to many academics and social commentators (including Wilkinson and Pickett, 2009), is that they are also the most unequal. Therefore, it would seem that some well-being metrics have a greater impact than others, with inequalities being the most important and others, such as being stuck in traffic, proving less so. Indeed, at best the more 'spurious' metrics such as this show confused logic; at worst they are an attempt to provide a smokescreen for real progressive thinking that will challenge the status quo (Tonybee, 2010).

Other commentators caution against this, arguing that the concept of happiness is too loosely applied as it can mean proportionate, progressive, deserved, unbiased or unprejudiced; that is, if the concept of fairness is to be more than a political buzzword, it needs to be grounded in an understanding of the relationships and organisations that make up a 'big society' (Norman, 2011). And it needs to be about genuine equality be it in terms of comparative wealth or gender parity, etc.

Furthermore, part of the problem lies in a misplaced faith in human rationality which has underpinned policy making for too long. To generate social mobility and reform a 'broken' society devoid of shared trust requires ensuring that people from different classes feel united in a common enterprise (Brooks, 2011). This needs to be built collaboratively on close-to-home issues.

Isolated examples of common concerns and joint action are of course not unusual. Take for instance the forward-thinking programme by Swedish councillors to establish a municipality free of fossil fuel and well adapted to climate change. A particularly novel aspect of efforts to stimulate positive behaviour change is the use of multimedia in the form of a short film to rally ordinary people to take action on climate change. (And part of an emerging trend around the world that also includes the use of comics in the USA and computer games in the UK e.g. Fate of the World, which puts computer game players at the helm of a future World Trade Organisation-style body with the task of saving the world by cutting carbon emissions: Vaughan, 2010.)

Common concerns and action on climate change in Växjö (Sweden)

Växjö is the capital and administrative centre of Kronoberg County and home to 83,000 residents. Over 60 per cent of the geographical area is characterised by vast forests and a multitude of lakes, which support local wood production businesses and a healthy tourism industry. Other major industries include information and new media technology, with Växjö acting as one of the media centres in Sweden where many television and radio programmes are produced.

Växjö's vision for the city is that it will operate and grow in a way that contributes to sustainable development, whereby the city's consumption and production are resource-effective and pollution free. Further to this, Växjö launched a bold campaign and commitment to be fossil fuel free by 2030, which involves both technical initiatives and behaviour change. Development towards a fossil-fuel-free Växjö has as its primary focus, energy security as well as environmental sustainability and local job creation. The programme contains a broad spectrum of activities aimed at improving energy efficiency and increasing the amount of energy from clean sources. Växjö has become a pioneer for its use of wooden biomass, as one of their main goals is for timber-based fuels to become the primary substitute for fossil fuels, thereby putting the surrounding forests to good use.

Other activities in the plan include biomass-based district heating and power generation, smaller-scale district heating, district cooling, biomass boilers for households, energy-efficient street lighting, solar panels, cycle paths and energy efficient buildings (World Clean Energy Awards, 2007).

Allied to these technological fixes has been a focus on educating and empowering residents to take action on a shared concern, that of low-carbon living. One novel way of advancing this has been to show in cinemas before the main film is screened a short film on how to live a low-carbon life style. Åsa Karlsson Björkmarker, Vice Chairman School and Childcare Board, Växjö comments:

> The film focused on individual consumption and CO2 emissions. We realised it was a complex issue, especially at the local governance level, given lots of

> these emissions arise from products we all buy that are sourced from overseas. So for instance, one of the issues we placed a particular emphasis on was resident's choices when it comes to transport.
>
> *(Monaghan, 2011a)*

Today 84 per cent of the city's heating comes from this type of fuel and, ten years after the campaign started, carbon dioxide emissions are down by 34 per cent per capita (compared to 2009 levels).

The shift to renewable fuels for heating, however, also has implications for economic sustainability, as the wood fuel is a locally managed and sustainable resource. The use of such local resources creates local employment opportunities and reduces the vulnerability associated with a dependence on outside supplies. Another benefit of switching to biomass for district heating has been that citizens are receiving comfortable, stable and clean heat for a low price.

The city's residents are also now benefiting from the next phase of the campaign which focuses on transport emissions; for example, Växjö has made the city centre a pedestrian zone by replacing pedestrian crossings with car crossings, set up a pro-cycling campaign targeting municipal employees and also offers a bicycle pool for employees at city hall. In addition to these cycling initiatives, Växjö is also looking at internal functions, specifically to help politicians look further than finances. Växjö now uses ecoBudget, an environmental budgeting model, for their financial accounting and business planning. The model is designed to show decision makers how to budget natural resources in the same way that they budget fiscal ones. Växjö has used ecoBudget for a coherent environmental management approach that supports their ambitions of becoming fossil-fuel free with clear time-related targets and cyclical evaluations of performance. Björkmarker concludes:

> Strong political consensus on strategy and scrutiny of performance has been key to our success on environmental issues to date. Our environmental programme is reviewed every five years and is approved by the local parliament. This requires good cross-political party working. We also conduct an Environmental Awareness Index survey every year to ensure the council is dealing with the right issues that affect our residents. The survey includes questions relating to people's consideration of the environmental consequences of their shopping, for instance whether they buy locally produced food. Overall, the trend is that residents are getting more aware, with the rating increasing to 64 per cent in 2010 from 63 per cent in 2009. Whilst we are satisfied with these results we must also recognise we need to do more. Firstly, transport remains a challenge and in particular the use of the car. We also want to help ensure more people take pride in and value our burgeoning reputation as the greenest city in Europe.
>
> *(Monaghan, 2011a)*

For further information visit http://www.vaxjo.se (accessed 19 May 2011).

So, as we can see from Växjö it is possible to bring people together around the single issue of climate change, albeit for different reasons. For some residents the close-to-home issue is energy security, whilst for others it is environmental sustainability, and for many it supports local jobs. What is less clear at the moment perhaps is how the use of novel social media is influencing people's behaviour and contributing to the council's goals.

A better understanding of how best to understand and influence behaviour change is the area we turn to next.

Significant behaviour or not?

There is a deep body of academic literature on so-called environmental significant behaviour (ESB), whereby behaviour is said to be influenced by psychological factors, aspects of physical and social context, personal capabilities and habits. This means to stimulate a certain behaviour requires controlling or shaping each of these factors. To understand this further, Stern (2000, 2005) proposes ESB can be divided into four categories:

- committed activism (e.g. volunteering with Greenpeace);
- environmental citizenship (e.g. donating money to Greenpeace);
- policy support and acceptance (e.g. writing to a politician to ask them to endorse Greenpeace's campaigns); and
- private-sphere environmental behaviours (e.g. buying sustainably forested wood products as recommended by Greenpeace).

Similarly, Guagnano et al. (1995) also recommend an 'A-B-C Model' which predicts that behaviour (B) is a monotonic function of attitudes (A) and external conditions (C). Which, returning to the former example, would mean that if a person believes that a Greenpeace campaign on climate change is important (A), with all their friends and family agreeing too and cost of petrol rising at the same time (C), then this may result in them taking part in the campaign's 'call to action' by for instance leaving their car at home from time to time and walking more (B). Importantly, the model also predicts a context in which the effectiveness of educational or information programmes intended to change attitudes (A), or regulatory and incentive programmes intended to change external conditions (C), may have less to do with the size of the intervention than with relative distributions of (A) and (C) in the population.

Whilst that is helpful for one issue on environmental sustainability in isolation, how could this be achieved across a number of values-based concepts?

The notion of a broader set of common causes and cultural values is a concept that has been picked up and developed by civil society actors such as WWF and Oxfam (Crompton, 2010). Crompton supports the idea that there is mounting evidence that facts play only a partial role in shaping people's judgement, and so there is an irony in much campaigning on humanitarian or environmental crises in that as society's understanding of the answers to these problems grows we tend to rely ever more heavily on data-rich tactics that may actually diminish public support for an enduring solution.

Crompton's research also concludes that people's values tend to cluster in remarkably similar ways across cultures. Common values here include the value placed on the sense of community, affiliation to friends and family, and self-development. At the same time, however, these values are also contingent upon the perceptions or 'gifts' of others which relate to social status and power, amongst other things. Critically, the former tend to trump the latter, and so it is these values of self-development, sense of community and affiliation to friends and family that must be activated and strengthened in order to nurture more successful societies.

Building on this point are the findings of a study by the National Centre for Research Methods at Southampton University (Ramesh, 2010b), which reported that poverty and gross inequality are six times more likely than ethnic diversity to cause British people to be suspicious of their neighbours. This repudiation to the argument that multicultural societies make people uneasy and less trusting of strangers is especially timely during a period of economic downturn when concerns over job security, often understandably, tend to surface. Some commentators also cite gang riots from London to Los Angeles as a startling example of when a lack of parity between citizens leads to societies breaking down.

Separate lives: lessons on riots and gangs from Los Angeles (USA)

The spate of inner city rioting and looting by youths from Manchester, Birmingham, London and other cities in England during August 2011 was a shocking case of people leading 'separate lives'; although people live in the same city they occupy different 'planets' depending on their class, wealth and education (Carrell et al., 2011). Therein, there is no sense of community solidarity.

In the immediate aftermath of these riots the government's main focus was on reforms to law and order on the basis that there was clear evidence that gangs had been at the centre of much of the violence that swept Britain's cities (Watt et al., 2011). Measures proposed at the time ranged from evictions from social housing, to withdrawal of welfare benefits for convicted rioters and to injunctions to prevent gangs organising. As well as the sheer horror and fear many people experienced, or the concerns they had about the police losing control of public order, there was the disturbing fact that 22 per cent of the thousands of rioters were children aged just 11–17.

According to Carrell et al. (2011), lessons learned from gang violence in the USA in the 1990s, such as in Los Angeles, show that reforms to policing tactics, whilst necessary, are not sufficient – only greater equality can nurture community cohesion. Preventing young boys and girls being recruited to the gangs and the rehabilitation of existing gang members was fundamentally important here. In terms of prevention, in addition to injunctions that prevent gangs organising, the Los Angeles police and community organised activities for children at risk of joining gangs, such as football matches and surf camps. With regard to rehabilitating existing gang members, the Los

Angeles scheme provided mentoring tailored to individual needs, with support on entering education and finding employment. Rehabilitated ex-gang members were also recruited to mentor current gang members to provide a legitimate voice to these initiatives. In short, Los Angeles' approach to intervention, without excusing any of the organised criminality, was as much about closing the inequality gap as it was about restoring law and order.

And so, ultimately, it seems the existence of successful society is dependent on how happy people are in terms of reaching their life's aspiration for themselves, their families and wider community.

Yet, to do so, this will surely require a power shift. That is, a devolution of power to citizens, and in particular the comparatively poor and disenfranchised. But just how successful has this local diffusion of power been in practice to date, in terms of both scale and longevity? And what, if anything, does this mean in terms of citizens taking on more responsibility as well as power?

Conclusions

- Revolutions throughout history have shared common aspirations for liberty, equality and representation.
- This shows us that human nature has a strong ability to effect change in society.
- Happiness and well-being are good measures of the success of a society, and addressing inequalities is the best route to achieving them.
- Achieving the desired change in behaviour requires understanding the array of attitudes and external factors that influence ordinary people.
- In doing so, we need to spend more time focusing on what brings people together as opposed to what sets them apart.

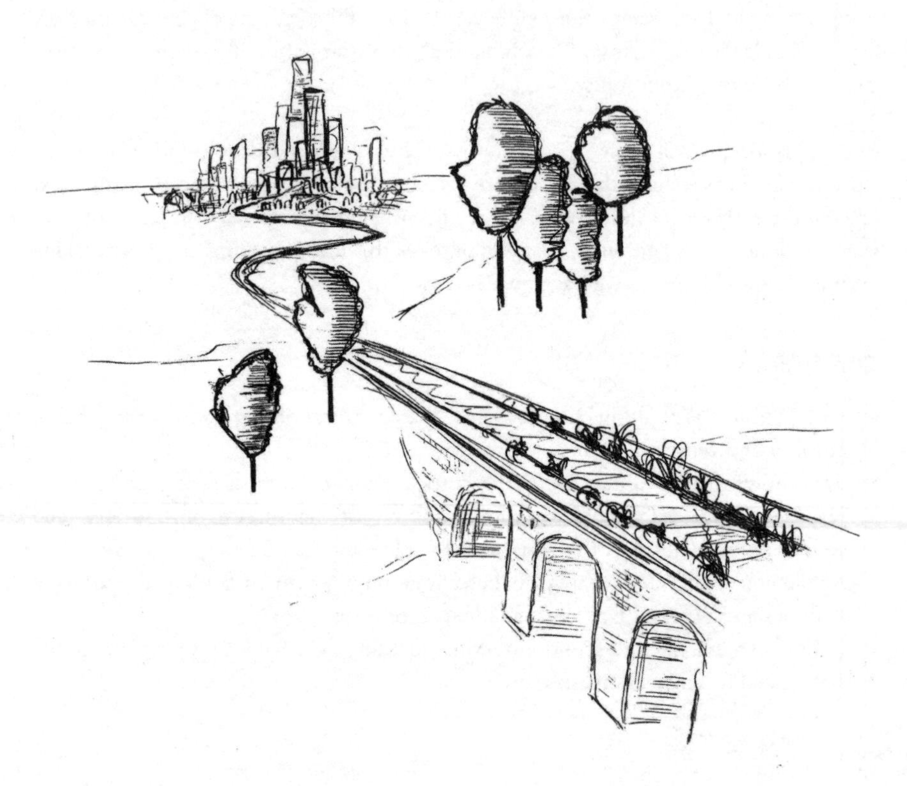

Aug 11.

Part Two
Localism without government

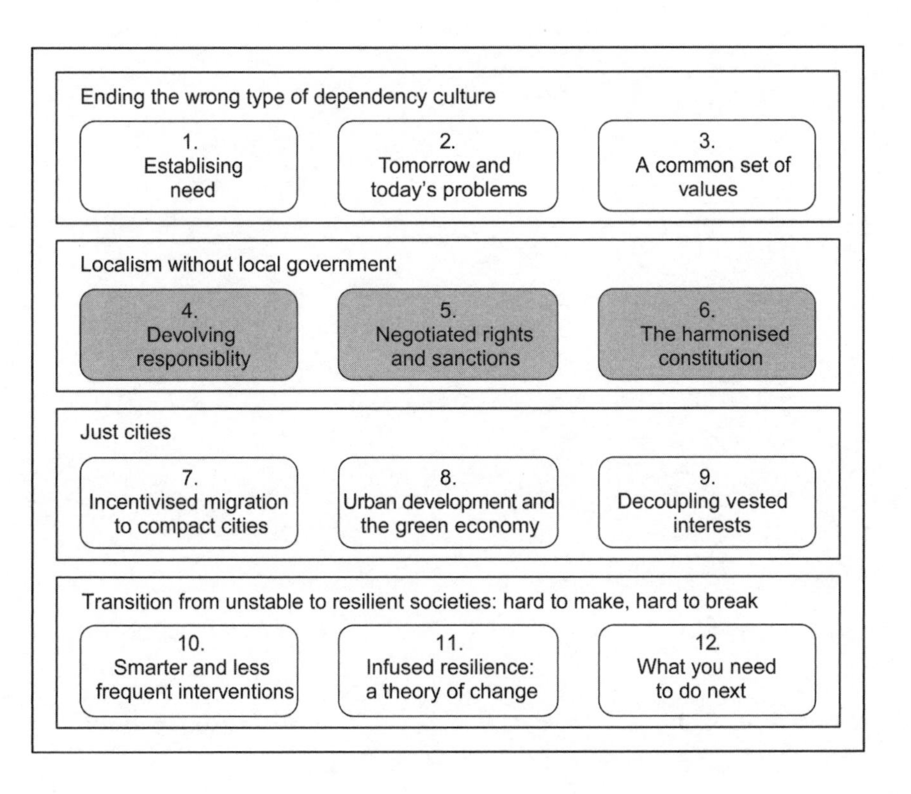

4

DEVOLVING RESPONSIBILITY

> The interface between local citizen-led action and representative democracy is right at the cutting edge of sustainable development.
>
> Halina Ward, CEO, Foundation for Democracy and Sustainable Development (Ward, 2009)

Communities on the front line or in the firing line?

Asking community groups or private sector contractors to take on front-line public services traditionally delivered by local government is now more popular than ever before, ranging from business-run prisons in the UK to volunteer managed libraries in the USA. Drivers range from a desperate or cynical way of making budget cuts in an age of austerity to political interest on both the right and left wings for a new way of doing business. Regardless of these drivers, or whether this transformation is underlined by a preference for smaller government or a desire for the co-production of solutions, the direction of travel is the same: outsourcing or commissioning public services such as libraries and street cleaning will continue to occur. Profit-making companies, cooperatives, social enterprises and the voluntary and community sector are all in line to take on these services. Regarding the latter, in the UK, local community groups have even been given the legal right to take over services as part of a 'big society' act of parliament.

The UK Localism Act 2011 gives community groups the right in law to protect, improve and even take on local front-line services ranging from children's centres, social care services and transport. Advocates of the new powers argue that it is needed because many local services from post offices to swimming pools have been closing down despite communities being willing and able to take them over (Curtis and Ramesh, 2010). Opponents of the Localism Act counter that it is potentially undemocratic as it passes power from elected councillors to unelected bodies (Jenkins, 2010).

Lesser government and bigger society in Suffolk (UK) and Havana (Cuba)

During 2010, Suffolk County Council responded to the Localism Act challenge by setting out an ambitious and controversial vision for setting up a 'virtual' council with the intention of outsourcing all public services to social enterprises, the voluntary sector or private companies (Bawden, 2010). The aim is to transform the council that provides most, if not all, local public services itself into what it termed an 'enabling' council that commissions services rather than delivers them. The intention is to make a 30 per cent saving on its US$1.8 million annual budget. An additional outcome of the plan is that just a few hundred people would retain direct employment by the council from a workforce of around 27,000 staff, with the remainder either being transferred to the winning contractor or losing their job. The primary driver for this new way of operating is the council coming to terms with the central government spending cuts in a new age of financial austerity (which will see a reduction in the size of public spending by US$133.4 billion over four years from 2011 onwards: Murray, 2011).

Somewhat similarly, also in 2010, one million public sector workers in Havana and across the country were told to become entrepreneurs in a bid to boost Cuba's private sector, on the basis it was no longer possible to protect and subsidise salaries (Carroll, 2010). Given that 85 per cent of the labour force work for the state (including taxi drivers and hairdressers), this is a gargantuan switch to privatisation by communist leaders, a notion backed up by a state study which reported that the change could be undermined if it did not act to address a lack of experience, insufficient skill levels and low initiative.

With the cases of Suffolk and Havana, regardless of bottom-line savings, one major question remains: how does such innovation in the decentralisation of services represent better *value for money* in terms of social outcomes? For instance, Doward and Stevens (2011) point out that whilst some studies show clear financial benefits to both service users and taxpayers from opening up the public sector to private companies (with cost savings of between 10 per cent and 30 per cent), the drive for the lowest-cost providers may prove to be counterproductive if this results in lower wages, job insecurity or job losses amongst the local workforce which then ratchets up welfare support needs. (It is worthwhile to note that, as this book goes to print, Suffolk's 'virtual council' programme is being abandoned because it has proven deeply unpopular with the public.)

Yet, despite worries raised about outsourcing or privatisation of activities in Suffolk and Havana in particular, there are other stories of success in challenging circumstances too, as witnessed through more 'bottom-up' community interventions from Venezuela, the USA and Nepal ranging from the fire service to resolving criminal disputes. A key differentiator here to the previous two examples appears to be that solutions are being proposed and led by ordinary people as opposed to 'top-down' ideas being imposed.

Bottom-up community provision in Caracas (Venezuela), Texas (USA) and Purena (Nepal)

Since 1998, the Venezuelan transfer of public services into worker-led enterprises has led to them now making up a major part of the 100,000 cooperatives that account for 18 per cent of the entire workforce (Mayo and Tizard, 2010). The government has facilitated this transfer by giving these new cooperatives access to credit and state procurement contracts and by offering technical support. One key component of this apparent success is that cooperatives were not seen by the state as an easy way of reducing costs in terms of shedding staff through outsourcing but rather as a better way of delivering quality public services.

Moreover, these new cooperatives often have lower overheads than the public sector and are frequently more innovative. Advocates including Mayo and Tizard emphasise the importance of ensuring, as Venezuela did, that such cooperatives respect the core principles of independence and open membership. To safeguard this, there needs to be an asset lock against demutualisation, as well as staff consent at all levels.

Another form of bottom-up community provision is seen in Plano, Texas, with volunteers who help to run the city's services, ranging from the fire service to teaching children about energy conservation (Williams, 2011). As of 2011, some 4,000 of the town's 280,000 residents are signed up to the 28-year-old Volunteers in Plano (VIP) programme (a network of local volunteers). The estimated value of this innovative programme is estimated at US$1.66 million per annum, with a cost of only US$150,000 to coordinate the scheme. Critical to this success, however, is the understanding that VIP's aim is to *enhance* services not to *replace* them. Thus, a major downside of this initiative is that volunteers are free to take time off as and when they want and as such cannot provide the necessary consistency in a way the paid staff can.

Similar approaches from the community have been seen in Nepal since 1997, through the creation of paralegal committees (PLC), whereby community representatives come together to address life and death matters themselves where the police or justice courts cannot or will not. On the basis that the majority of the country's disputes are resolved by informal processes for the poor and excluded, the women's paralegal committee (PLC) in Purena, a predominantly Muslim village in mid-west Nepal, consists of a group of courageous volunteers who visit homes to mediate in cases of domestic abuse, sexual harassment and child marriage (Brunt, 2010). The benefit this brings is a collective agreement about what is acceptable and what is not, whereas otherwise justice would not be served. Any reliance on non-state mechanisms is only possible though NGOs (such as Plan Nepal) supporting legal training and having formal access to state justice mechanisms in the event that mediation fails.

These stories illustrate the diversity of public service duties that different parts of a society may or may not be willing and able to take on, depending on particular circumstances. Indeed they may even be better placed than statutory authorities to deliver these in the first place. Factors influencing the ability and appetite for this range from the service in question being of common interest to community members, to whether there is local capacity to help, support sources on tap to nurture or coordinate such enthusiasm and, ultimately, legal imperative for citizens to take on new responsibilities.

Yet, when and how these factors apply is not always clear from the outset and problems can arise both around people's own perceptions of their abilities and what they can do to help (even if they are willing), as well as, sometimes, an under appreciation by the authorities of ordinary people's competency to do the right thing, even in times of duress. Regarding the former, a good example of this is survey data by IEMA (Environmentalist, 2011a), which shows that, while 73 per cent of the UK general public confirmed the environment was important to them, only 45 per cent said they believed there was something they could do personally to protect it. A common misperception amongst government emergency planners, however, is that, if involved in a major catastrophe, such as a tsunami or terrorist attack, most ordinary people cannot be trusted as they will panic from the onset of fear. Consequently, top-down approaches run by technocrats may tend to automatically discount any helpful contribution from the public. Yet there is a large body of research – from the Second World War bombing of Hamburg to Al-Qaida atrocities on 11 September 2001 – that shows the opposite is often the case (Furedi, 2004). The greatest waste, according to Furedi, is the billions of dollars invested in civic contingencies instead of harnessing the natural abilities of ordinary people to build community resilience, which will do *more* to protect society.

By extension, the value of a clearer interpretation of people's abilities, their right to act and also their responsibilities in taking on new power is imperative, starting with understanding why, for some, community empowerment is a panacea for societal brittleness ranging from low enterprise to social conflict.

Navigating shifts in power, rights and responsibilities

The very term 'empowerment', its definitions and whether it is solely about the relationship between communities and public services or is also about wider activities taking place within communities has long been considered by social researchers and commentators (Chanan, 2009). In turn, desired outcomes from such empowerment vary too, ranging from community cohesion, an increased sense of belonging to neighbourhoods and equality of civic participation, to regular and valued volunteering and a thriving third sector.

This in practice may simply mean cleaner streets, lower crime or higher property values for some. For others, it is about ever increasing aspirational goals such as reversing democratic disengagement and the disconnection of citizens from the state (Prendergast, 2008). And, thus, appropriate solutions are contemplated for the perceived schism between *representative* and *participatory* democracy, a failure to transfer power in a meaningful way and a lack of clarity and transparency in the accountability of decision making. A great example of getting this right is the involvement of marginalised and impoverished local women in the reconstruction of their tsunami devastated communities in Sri Lanka (Monaghan, 2011b).

Centring women in governance in Moratuwa and Matara (Sri Lanka)

Moratuwa is a city on the south-western coast of Sri Lanka, situated 18 kilometres south of the capital, Colombo. Moratuwa is surrounded by water on three sides, by the Indian Ocean on the west and the Bolgoda lake on the south and east. The city has a population of over 177,000.

Matara is a city on the southern coast of Sri Lanka, 160 kilometres from Colombo. The city has a population of over 76,000. City industries include the manufacture of furniture, rubber products, batteries, transformers and wood handicrafts as well as fishing trade centres.

With Sri Lanka being one of the countries worst affected by the 2004 tsunami disaster – the Government of Sri Lanka estimated that reconstruction costs would be approximately US$2 billion (ICSC, 2009) – both cities were devastated. One of the major social impacts of the tsunami was the way in which it, and responses to it, affected men and women differently. According to the International Centre for Sustainable Cities (ICSC), a Canadian NGO, even prior to the tsunami Sri Lankan women were often inhibited from reaching their potential because of responsibilities to their family and problems arising from women having low levels of education and skills, and having a low social status in their community. These factors exacerbated the problems that arose in the immediate aftermath of the tsunami. Consequently, to help address this dilemma the ICSC and Sevanatha Urban Resources Centre, a Sri Lankan organisation, led the Centering Women in Reconstruction and Governance Project (CWRG) between 2006 and 2009.

The CWRG (known locally as Viru Vanitha which means 'powerful women') was designed to address:

- weak communication between appropriate levels of government, NGOs and grassroots communities;
- the continuation of a dependency syndrome by the local community post tsunami;
- poor distribution of aid to impoverished communities near Columbo (compared to the east and west of Sri Lanka);
- an absence of participatory approaches in reconstruction;
- a lack of NGO and local government involvement of women.

The intended outcomes for the project were:

- to improve women's participation in decision making and the local council's responsiveness to their needs;

- to equip women with income generation skills for livelihoods that enable them to sustain their families and communities;
- to reconstruct the post tsunami environment.

Jane McRae, CEO of ICSC, says:

> Critically for us in scoping the project were the facts that it built on 'urban greening' work here and so there was both a track record of need, and of success, and that the new activities followed positive developments in similar situations in Turkey and India.
>
> *(Monaghan, 2011b)*

To support these project outcomes, community resource centres (CRCs) were built in the two cities of Moratuwa and Matara. The goal of the CRCs, which were run by women, was to establish a safe place for the communities to organise, to exchange information, to receive or provide livelihood training and education, to provide crucial support services and to sustain women's public participation through and beyond the tsunami recovery period.

Sumana Wijerathne, Project Manager at the ICSC, states:

> The big success to date has been around the community contracting procedure for the reconstruction. Women have mobilised to run the process from the initial terms of reference to the end use. Key to this has been the opening of the well connected and resourced Women's Bank. This has provided a lasting legacy in terms of not just the physical building, but new financial education and the confidence to challenge male counterparts in public.
>
> *(Monaghan, 2011b)*

However, Wijerathne adds that the project was so successful it had additional outcomes:

> It did something that went beyond the original scope of the project in the form of problem solving skills that could be applied elsewhere, as evidenced by the Mayor being so impressed he recommended to the national government that the women be given housing loans to distribute locally too and the municipal authority establishing a Gender Council and Women's Community Fund, demonstrating the competency of and respect in their decision-making.
>
> *(Monaghan, 2011b)*

Samantha Anderson, Senior Project Manager Asia, ICSC, concludes:

> New and transferable learning from this success is, firstly, further evidence of the added value of focusing investment on community-led, not central government-led capacity building, and secondly, the need for post-disaster aid to

> focus on male as well as female education in order to change male attitudes to female empowerment in order to sustain any positive impacts.
>
> *(Monaghan, 2011b)*

For further information visit http://sustainablecities.net (accessed 25 May 2011).

A key strength of this empowerment initiative was how 'success breeds more success', in terms of the women having the legitimacy to take on more responsibilities from male elders as they triumphed with earlier duties. A limitation of the transformation programme was the absence of education work to shift attitudes of men from the outset that empowering women was good news for their community.

It is this very human collaboration, of people working together for a common aim or cause, which is seen by many social commentators to be the primary pathway for progress on empowerment. Given such interaction either acts as the communal glue that connects people together (Rowson et al., 2010) or is the fountain of all good ideas (Johnson, 2010).

However, this optimism is tempered by calls for a greater appreciation of the limits to the size a human social group can realistically be maintained in order to have a real sense of community. For instance, Dunbar, (2011) proposes that the perfect number is 150.

Ostrom's (1990) acclaimed findings on communally managed resources also analysed the reasons for success and failure of empowerment, but take a somewhat different approach where emphasis is placed on the absolute need to negotiate sanctions, as well as rights, and there is a penalty for members of the community who fail to meet their responsibilities.

So, can such subtleties in the governance of these rights and wrongs be dealt with? This is something we will look at in Chapter 5.

Conclusions

- The rationale for moving to smaller government and bigger society may vary, from the need to make budget cuts, to a desire to empower communities.
- There is a mismatch of people's perceptions of their ability to contribute and there can be a lack of understanding of how valuable this contribution could be by those in authority.
- Any devolving of responsibility to communities should be done in a way that is robust in terms of nurturing positive social outcomes that can be sustained in the long term.
- Human collaboration can bring people together around a common cause. To sustain such new forms of organisation may eventually require formalising how communities govern themselves, including oversight of rights and wrongs.

5

NEGOTIATED RIGHTS AND SANCTIONS

> We take great pride in our 'womb to tomb' healthcare ... at the same time the people's contribution to climate resilience is a key focus for a region visited by several natural disasters every year.
>
> Violeta Somera Seva, Senior Adviser, Makati City, Philippines (Seva, 2011)

Return to fairness through contribution

To help make change universally desirable and unite people around a common set of values, as discussed in earlier chapters, a sense of fairness needs to pervade any and all strategies for creating sustainable societies. Fairness fosters trust and democracy and these, in turn foster, openness and reciprocity within society, institutions and markets (Hutton, 2010).

However, fairness is a very over-used concept. For genuine fairness to apply a 'contributory principle' is essential, which means that people enjoy benefits from a system proportionate to those they have given to the same system. Some examples of this are controversial, such as a UK citizen's welfare claims becoming linked more closely to what they paid into the system (Wintour, 2011) – that is, relying less on payments and moving towards job guarantees and access to housing, which mirrors an Australian scheme that forces unemployed people to take temporary work or have their benefits cut. The intention is that participation will develop people's skills and networks and should assist them in securing a permanent job (Christie, 2011). By doing so, it is argued by IPPR, amongst others, that this type of intervention binds people together around one of the things (in addition to love, prosperity and so on) that matter most: responsibility. In IPPR's case, one policy proposal is for the UK government to establish a 'national salary insurance' payment scheme as part of reforming the welfare state, as detailed further in Box 5.1 (Cooke, 2011). The argument for such a scheme is especially relevant in the current age of unprecedented financial austerity *and* great social and environmental ills, with many people angry about what they perceive as 'not everyone doing their bit' yet not worse off (e.g. if they are able to work but never actively seek employment and yet still receive state benefits), or indeed sometimes better off than those who were contributing (e.g. if they are 'benefit cheats' who claim state support but are also informal workers who do not pay tax).

Box 5.1 Reforming the welfare state to provide salary insurance

A study by IPPR (Cooke, 2011) suggests that the UK welfare state has been undermined through a perception that it is not demanding enough of people who *do not work* and, simultaneously, not protective enough of those who *do* work.

To address this problem the IPPR proposes that the UK government should establish National Salary Insurance (NSI) which would provide people with payments if they lose their job that would subsequently be repaid when they return to work. An NSI would offer working people who become unemployed up to 70 per cent of their previous earnings for up to half a year (capped at a maximum amount per week).

The policy is designed so that the government recovers the total cash cost of higher benefit payments. Therefore, implementing the scheme during a recession, when unemployment is higher and public resources are scarce, is not a barrier to act.

This 'contributory principle' interpretation of fairness holds just as true for any rights or sanctions-based approaches to development, both in terms of how the approach is agreed as well as the problems it addresses. That is, it needs to be negotiated.

Insights from Ostrom (1990) on so-called Community Pool Resources (CPR) – communally managed meadowlands, forests, fisheries and other assets – may be extremely valuable here. Based on learning from Japan, Spain and Switzerland, CPRs generally involve local institutions performing collective work to enhance and maintain the yield of the area (such as annual burning or cutting of timber to thatch in the case of meadowlands) with each household having an obligation to contribute a share to such efforts. If any household fails to perform their share of work, fines are imposed by the local village institution (which could, for example be a donation to the village school). According to Ostrom, the long-term success of these locally designed rule systems indicates that it is not necessary for external regulation of the CPRs to be imposed coercively by national government, as local 'self-policing' works well. Given this, Ostrom illustrates a number of key design principles illustrated by long-enduring CPR institutions:

- Clearly defined boundaries – individuals or households who have rights to withdraw resource from the CPR must be defined, as must the boundaries of the CPR itself.
- Alignment between rules of appropriation and provision and local conditions (i.e. appropriation rules restricting time, place, technology or quantity of resource

are related to local conditions and to provision rules requiring labour, material or money).

- Collective-choice arrangements whereby most individuals affected by the operational rules can participate in modifying the rules.
- Designated monitors, who actively audit CPR conditions and appropriate behaviour, are accountable to the appropriators or are the appropriators.
- Appropriators who violate operational rules are likely to be assigned graduated sanctions, depending on the seriousness and context of the offence, by other appropriators, by officials accountable to the appropriators, or both.
- Appropriators and their officials have rapid access to low-cost arenas to resolve conflicts. The rights held by appropriators to devise and organise their own institutions are not challenged by national government authorities.

As this approach to collaborative governance has worked in a number of municipalities around the world, it would appear to lend itself to being replicated elsewhere (Monaghan, 2010). In accepting that CPR is valuable, is it therefore possible to scale-up from a single resource to a broader area-based approach for city level solutions?

Area-based negotiations

There is a wealth of learning as to how municipal authorities are attempting to renegotiate the terms of the relationship between themselves and their citizens, as well as between the citizens themselves. One area is reward or punishment schemes, often based on 'nudge theory' developed by academics Thaler and Sunstein – the idea that people do not automatically do the right thing but will respond if the best option is highlighted – which increasingly is seen as one valuable way to get community members to 'do the right thing' (Williams, 2010).

'Carrot and stick' citizenship schemes in Nestaudt an der Weinstrasse (Germany) and Huyton (UK)

The citizens of Nestaudt an der Weinstrasse are the most prolific recyclers in Germany, with 70 per cent of household waste recycled. Over the past three decades, the town has developed its approach, including education on waste management and in 2006 they also introduced financial incentives associated with recycling (Hickman, 2011). Citizens are not charged for any recycled waste they leave out, but all other waste which has to be incinerated is chargeable. This fee is made visible by being separated from the local tax bill, which increases awareness and visibility of the charges and, subsequently, waste costs arising for the council are lower now than when they started the scheme.

In Huyton, motorists caught using their mobile telephones (which is illegal in the UK) have to explain themselves to school children as part of their punishment (Liverpool Daily Post, 2010). Culprits caught while driving near a particular area are offered the choice of either penalty points on their driving licence and a cash fine or endure a grilling by pupils at the school. The drivers also have to watch a DVD of parents who have lost children in accidents where drivers were using mobile phones and undertake a simulation showing the dangers of using mobile telephones whilst driving.

Another avenue is novel forms of efforts to mobilise excluded groups and involve them in collective decision-making and action.

Mobilising excluded groups in planning and campaigning in San Carlos (USA) and London (UK)

A growing number of Californian cities are involving young people in planning processes, as officials discover the value of young people's insights on land use (Amsler, 2010). By engaging young people significant benefits are attained in the form of improved plans, leadership opportunities, citizenship and service, optimal use of youth-serving resources and exposure to careers in local government. Specifically, in the case of the city of San Carlos, this has been embedded to the extent that since 2006 student members of the city's Youth Advisory Council serve as rotating members of a voting position on the General Plan Advisory Committee, taking decisions on community spaces, housing opportunities and community life.

London Citizens is a campaign founded in 2001 by religious, trade union and community leaders in east London angry that family life had become impossible for many parents struggling to hold down jobs, commuting long distances or working anti-social hours (Stewart and Loweth, 2011). In response it mobilised citizens to challenge employers to voluntarily pay a 'living wage' – a mark-up on the legal required minimum wage – in order to improve people's quality of life. For over a decade London citizens have taken on the banks and persuaded schools, hospitals and shopping centres to raise pay for workers. Signatories include corporate giants such as Barclays, HSBC and KPMG.

A further instance of negotiated responsibilities is in the Basque Country where businesses are leading by example, or are being asked to 'do their bit'.

Economic and democratic renewal through enterprise in the Basque Country (Spain)

The once depressed Basque Country has been developed through democracy at work and social clubs in Bilbao, helping the city to thrive despite the current recession (Ramesh, 2011). With a jobless rate at 10 per cent whilst the national average is twice as much, the region gives an inspirational example of the value of cooperatives.

Support for this way of doing business is further reflected in the fact that Spain's constitution explicitly recognises that cooperatives encourage democracy by bringing workers together for the common good and so they consequently pay less tax (which, in turn, allows these cooperative enterprises to invest 10 per cent of their annual surpluses in the local community).

In addition to promoting democracy, the diverse and flexible nature of governance means the cooperative sector is more resilient to economic recession, say cooperative exponents such as Mondragón. So when the recession meant that the cooperative retail giant Mondragón in 2009 lost 14,000 jobs and cut pay by 8 per cent there was no worker dispute. Mondragón cannot make members redundant or sell off divisions, and so employees are guaranteed jobs for life. Losses in one division of Mondragón are offset by gains in other parts, and workers from failing cooperatives are redeployed to more successful cooperatives (Ramesh, 2011).

In comparison, it is interesting to note how one local government association in the UK has chastised fruit products company Del Monte for its allegedly irresponsible packaging policy whereby individually wrapped bananas were sold in convenience locations and contributed to the twin problems of waste landfill and litter (Poulter, 2011).

Of course, asking everyone – citizens, employees or companies – to 'do their bit' includes public officials too. An extreme and eye opening example comes from the city of Hebei in China, where the policy is for officials who neglect their families not to receive promotion.

No promotion for local officials who do not care for their family in Hebei (China)

Party leaders in the Hebei province of China refuse to promote officials unless they show sufficient care to their family, in the hope that it will set a good example to other citizens. The underpinning theory is that they cannot possibly do their jobs well if they do not treat their own families properly (Branigan, 2011). This policy even involves assessors quizzing spouses, parents and in-laws about how well they are treated, with complaints from unhappy relatives being a key factor in vetoing promotions.

A more telling example, though, is the city of Brighton and Hove, which has an ambitious vision to bring together a number of smart measures to control and influence behaviour together as part of boosting its biosphere reserve (Monaghan, 2011c).

Towards a collaborative contract to preserve the biosphere in Brighton and Hove (UK)

Located on the picturesque coast of south-east England, close to the South Downs National Park, and only a one-hour commute from London, the city of Brighton and Hove is one of the UK's preferred residential and visitor destinations. With a population of nearly half a million people it has a high density of businesses involved in media, particularly digital or new media, and since the 1990s has often been referred to as 'Silicon Beach'.

Aware of the importance of the local environment to its economy (e.g. tourism) and quality of life (e.g. leisure), in 2008 Brighton and Hove Council launched a campaign to become internationally recognised by the UN as a 'biosphere reserve' (Tibbetts, 2008). Biosphere reserves are areas of terrestrial and coastal ecosystems promoting solutions to reconcile the conservation of biodiversity with its sustainable use. Each biosphere reserve is intended to fulfil one or more of three basic functions: conservation, human development or education.

Biospheres are internationally recognised, but are nominated by national governments and remain under sovereign jurisdiction of the state where they are located (UNESCO, 2011). Research by Stanvliet and Parnell (2006) suggests that biosphere reserves can contribute to urban resilience by providing a valuable tool in the planning and management of more sustainable cities.

The project proposal was first debated in 2008 and by 2011 all the main political parties had included a commitment to the campaign in their local election manifestos, with the Green Party subsequently taking control of Brighton and Hove Council (becoming the first council in UK history to be led by them). 'The Council's leadership is really committed to delivering the biosphere reserve as part of a wider and challenging vision for Brighton and Hove to become the greenest city in the country,' says Thurstan Crockett, the city council's Head of Sustainability and Environmental Policy (Monaghan, 2011c).

The chalk hills of the South Downs National Park have been identified as a biosphere reserve as the chalk grassland is an internationally rare habitat which is important for biodiversity and the management of woodland. The geology and topography of the area is important in supporting wildlife which is special to a chalk landscape, offering construction materials, and providing clean water. Crucially, the landscape and wildlife provides a green lung for the health of the city's residents and its soils form the basis of the agricultural land for producing food, much of which is increasingly consumed locally.

The sea is another important natural resource for the area. It has a rich ecology, is important for recreation and as a source of food and other services that the local population rely upon. This is in addition, of course, to attracting tourists.

The Council believes Brighton and Hove is well placed for designation as a biosphere reserve because:

- the natural environment is easily accessible with the city bordered by both the South Downs National Park and the sea;
- the city has a strong social history of environmental activism and innovation – locally, nationally and also at an international level. It played a leading role in the campaign to get the South Downs designated as a National Park; it was the first city to sign up to the 10:10 climate campaign; it was in the top three in the Forum for the Future's National Sustainable Cities Index for the past four years (see Forum for the Future, 2010); it has gained the World Health Organisation's 'Healthy City' designation;
- there is a strong culture of working partnerships within the city between all sectors (business, community and voluntary sector, local authority and public sector). The

city was one of the first areas in the country to establish a Local Strategic Partnership (LSP) and the recently launched Sustainable Community Strategy is committed to supporting the establishment of an urban biosphere reserve;
- the city is a centre of learning, with both the University of Sussex and Brighton University each housing well-established environmental and social departments that could provide a strong research, monitoring and reporting base for the development of the biosphere reserve;
- 5,000 hectares of downland, extending beyond the city limits, are in public ownership. This offers a rare opportunity for the city to work in partnership with its tenant farmers and others to influence farming, to improve access and understanding, to conserve and enhance natural habitats, and to better integrate farming into the life of the city.

In terms of what a biosphere reserve brings to Brighton and Hove, the council emphasises that research has repeatedly shown that green cities provide health benefits and attract investment. Similarly, surveys by the council have shown that people in the city highly value the good access to green space, the National Park, the coastline and the sea. Against these clear benefits, Brighton and Hove acknowledges it must accommodate potentially large numbers of new housing units without significantly extending its boundary and there is a risk that the green character of the city could be damaged. So, designation as a biosphere reserve would further support recognition for the importance of the natural environment to the city as well as attract the resources needed to find new ways to unite the economic and environmental issues facing the city.

Crockett continues:

> The council has an ambition to create a sustainable future for the city's communities based on operating within our environmental limits. This is a major transformation, requiring serious investment, but we are confident the return on this investment will be well worth it. UNESCO designation can aid this, as our analysis indicates that this leverages external funds against the initial outlay at a ratio of 8:1. We also know that the local utility company Southern Water spends US$8m per annum treating water so it is drinkable, yet this is processed through our own agricultural estate; so by bringing management of the South Down National Park back 'in house' we believe we can avoid the costs of treating or importing water, enhance water security as well as realise wider benefits such as promoting responsible tourism. This is why the council has been able to agree the business case for establishing a new post to coordinate this initiative going forward.
>
> *(Monaghan, 2011c)*

By greening the city in this way, biosphere reserve designation would specifically lead to:

- transition to a low-carbon economy through the promotion of responsible tourism and attracting more green businesses;

- increasing levels of exercise, improved health and well-being, aided by high-quality green space and improved access for pedestrians, cyclists and public transport users;
- improved diet through increased consumption of locally grown fresh food;
- reduced air pollution through the filtering effect of vegetation (and through encouraging more people to walk and cycle);
- a city better prepared for the effects of climate change through reducing storm water runoff (increased green/porous surfaces); enhancing aquifer replenishment (increased porous surfaces); reducing water demand (greater rainwater recycling); more shade in the summer (more urban trees); cleaner seas (reducing risk of storm water overflows); increasing the number of innovative micro-renewable energy projects (at both domestic and industrial sites); and increased local food production and consumption (in and around the city) to reduce food miles;
- reduced carbon emissions through reducing the need for air conditioning (increasing summer shading and adapting building design); encouraging more people to walk, cycle and use public transport; encouraging greater local use of the city's green space (reducing need to travel for high-quality experience); developing more energy-efficient buildings (both new and existing) and by reducing domestic and industrial energy consumption;
- improving the way the council's 5,000 ha downland estate is managed. It would be seen as a centrally important asset with multiple uses and potential sources of income, compatible with its National Park status. This will ensure the vital ecosystem services that are provided by this area, including clean drinking water, are better recognised and delivered to a high standard.

The council appreciates that a new approach to governance will be vital to success, which means working through the LSP with business representatives, emergency services, government agencies, NGOs and other local councils, as well as elected members. Crockett concludes:

> A working group of the LSP will be accountable for delivering against obligations to UNESCO. Our revised Local Development Framework in 2012 will set out any new powers that restrict or encourage a change in land use by residents, visitors or developers. We expect it to be tremendously challenging but very rewarding over the next 3 years.

For further information visit http://www.brighton-hove.gov.uk (accessed 22 July 2011).

The outstanding feature of Brighton and Hove's story is how all the main political parties are strongly committed to 'redrawing the lines' of what should no longer be allowed in order to preserve their much-valued local environment. Just as critically, there is broad consensus as to how this should be done. What is less clear at this stage, however, is how any new controls will be administered, for instance if developers breach new planning guidelines and, indeed, whether the council has enough powers and resources to take such punitive action.

These impressive stories from the Basque Country and Brighton and Hove also appear to provide evidence of the need to have an enabling framework at the national level which permits greater freedom at the local level. It is this blending of local and state level constitutions which is the focus of Chapter 6.

Conclusions

- Everyone needs to contribute – this is the fairest way. Carrot and stick options are available to support the move to all contributing fairly, and each can work depending on the issue.
- Fairness applies universally – to citizens, business and local authorities – and involvement fosters social engagement and strengthens democracy.
- To work properly, however, there need to be locally negotiated rights and sanctions.

6

THE HARMONISED CONSTITUTION

> It makes world history. Earth is the mother of all ... It establishes a new relationship between man and nature, the harmony of which must be preserved as a guarantee of the quality of its regeneration.
>
> Alvaro García Linera, Bolivia's Vice President (after announcing the country's intention to pass laws granting nature equal rights to humans) (Vidal, 2011a)

The journey from rights to responsibilities to subsidiarity

National constitutions or bills of rights have shown great symbolic and cultural importance, across the globe and often in the face of grave threats and challenges to a country's stability and system of values. From Magna Carta in 1215 to the constitution of the Republic of South Africa in 1996, each country's history illustrates the proud traditions or modern revolutions of liberty upon which current frameworks of democratic rights are built (Ministry of Justice, 2009). Yet, responsibilities have generally not been afforded the same prominence as rights within constitutional architecture, despite the fact that many duties often already exist in statute or common law (or their equivalents).

The notion of the fundamental link between rights and responsibilities, however, is not new, either in theory or practice, says the Ministry of Justice (2009). Rather, there is a twofold challenge: firstly, how best to remind people of the importance of individual responsibility and to give it greater prominence, and, secondly, to ensure that national constitutions or ethical policy frameworks both empower and enable local communities to take action on the ground, and that this is also extended to include public authorities and corporations alike.

Regarding the first challenge of giving more prominence to individual responsibility, for instance, the Ministry of Justice (2009) has proposed that responsibilities could include: treating health service and other public service staff with respect; safeguarding and promoting the well-being of children and other vulnerable people in our care; living within our environmental limits; participating in civic society through voting and jury service; assisting the police in reporting crimes and cooperation with the prosecution agencies; paying taxes and obeying the law. However, to date, the UK Government has decided against the introduction of a general model of either directly enforceable legal rights or responsibilities.

This is on the basis that, specifically in the case of responsibilities, the Government believes the imposition of new penalties is unlikely to be the best way to garner a sense of civic duty or encourage respect and tolerance for fellow citizens or greater voter turnout. In saying this the Government also acknowledges that a counter argument to this stance states that an over-emphasis on rights, to the exclusion of notions of responsibility, can lead to a 'me' society rather than a 'we' society, in which an exclusive focus on one's own individual rights and liberties would always overtake concerns of our collective security and wellbeing.

Tellingly, other countries have taken a very different route through their constitution, bill of rights or other national instruments that do. Where this double emphasis occurs, the dependency between the two is often underlined – examples include the USA, South Africa, Poland and Australia (Ministry of Justice, 2009). Each is detailed here:

- The American Declaration of Rights and Duties of Man 1948 states 'The fulfilment of duty by each individual is a prerequisite to the rights of all. Rights and duties are interrelated in every social and political activity of man. While rights exalt individual liberty, duties express the dignity of that liberty'.
- The African Charter on Human and Peoples' Rights states 'Every individual shall have duties towards his family and society, the State and other legally recognised communities and the international community' and that 'The rights and freedoms of each individual shall be exercised with due regard to the rights of others, collective security, morality and common interest' and also that 'Each individual shall have the duty to respect and consider his fellow beings without discrimination, and to maintain relations aimed at promoting, safeguarding and reinforcing mutual respect and tolerance'.
- The Polish constitution states that 'Loyalty to the Republic of Poland, as well as concern for the common good, shall be the duty of every Polish citizen' and that 'Everyone should observe the law of the Republic of Poland'.
- The Australian Citizenship Act 2007 states that 'Parliament recognises that persons conferred Australian citizenship enjoy these rights and undertake to accept these obligations: by pledging loyalty to Australia and its people; and by sharing their democratic beliefs; and by respecting their rights and liberties; and by upholding and obeying the laws of Australia'.

More recently, it is also possible to see more modern dilemmas being played out in terms of the new interpretations or upgrades of existing constitutions. For example, immigration authorities in France refused to grant nationality to applicants whose negative attitude toward women were considered incompatible with the values of the republic in respect of the equality of men and women (Willsher, 2011); and policy makers in Iceland are drawing up a new constitution through social media introducing new checks and responsibilities for citizens and parliament to prevent a repeat of the 2008/09 financial crisis (Siddique, 2011).

In respect of the second challenge of enabling local action, United Cities and Local Governments (UCLG, 2007) reminds us that, in a world where more than half of humanity lives in cities, local authorities are vital to democratic renewal and a sense of citizenship. According to UCLG, for good local governance here, the principle of 'subsidiarity' is key, whereby decisions should be made at the level of government closest to citizens.

Yet, according to UCLG (2007), whilst in the past two decades decentralisation has become the 'norm' in most countries around the world, many reforms are either very recent or are facing difficulties in their implementation. For example, in North America, central governments have devolved more of the responsibility for financing activities to the local level and, whilst this has meant less fiscal support from above, it has also resulted in fewer regulatory restrictions from above, which has led to new modes of service delivery through privatisation and public–private partnerships (PPPs). In Africa, the UCLG states that the implementation of decentralisation has rarely been well planned, and in Senegal and Burkina Faso there is no real plan for implementation at all. In the Middle East, in spite of political, religious and military tensions, there have been some notable advances (in particular the first local elections in Saudi Arabia).

According to UCLG (2007), issues of particular concern are financing and staff, especially in the countries of the South; that is, the impact on local autonomy when the level of financial control is deficient or non-existent given the tendency of central government to absorb a larger share of resources, or if the municipal authority does not have adequately skilled personnel to handle decentralisation and enable local communities to improve access to and the quality of services.

In Africa, for instance, while the public assertion of new nominal powers of local government is widespread, the transfer of real executive power is still rare. As a result, many countries are in an active tug of war over the state of decentralised democracy. A notable example of this is in South Africa, where the end of Apartheid saw a new approach to governance emerge based on decentralisation and public participation (UCLG, 2007). An excellent example of this is from Cape Town, South Africa (Monaghan, 2011d).

A constructive constitution for Cape Town (South Africa)

The second largest city in South Africa, with a population of 3.5 million people, Cape Town is the country's legislative capital where the national parliament is found. Located on the southern tip, it is a popular tourist destination with its famous harbour and mountain scenery. In addition to the tourist industry, and public sector employment, information technology is a major and growing industry for the city.

According to the council, the city of Cape Town strives to be at the forefront of sustainable development within South African local government. One of the aims of the city is to establish an awareness of the physical limits of resource use (e.g. electricity) on future development, to consequently minimise the impacts of resource use on the environment and to enhance people's quality of life.

Consequently, in 2006 Cape Town developed an energy and climate change strategy to integrate sustainable energy approaches into its core functions, with a framework that provides a clear vision and direction for the city as a whole. An adaptation framework has been produced in response to the potential short to medium-term impacts of climate change in the Cape metropolitan area. As part of this various projects and programmes addressing energy and climate change are run by the Environmental Resource Management Department, including:

- the *Kuyasa Energy Efficiency Project*, which retrofits existing low-income with solar water heaters and improved ceiling insulation measures;
- the Solar Water Heater Advancement Programme, which aims to encourage and facilitate the adoption of solar water heaters in Cape Town.

As a result of this Cape Town has become a centre for sustainability thinking. Yet, at the same time, complex socio-institutional dynamics are constraining its transition towards becoming a sustainability city. Important changes are hampered by an unsupportive central government on fossil fuel-reliant energy sources, and the legacy of Apartheid-era spatial planning compounded by more recent urban sprawl, vulnerability to water scarcity and sea level rise as a result of climate change (UN-Habitat, 2011).

Given this, to help develop a number of strategies to integrate sustainable thinking into city growth plans, the Climate Change Think Tank has been established, representing a number of high-profile institutions and thought-leaders such as the Centre for African Cities (Cartwright, 2012).

Anton Cartwright, a researcher at Climate Change CityLab, African Centre for Cities, said:

> Despite the big challenges it faces, what the City of Cape Town does have in its favour is institutional capacity to problem solve through a well functioning local government along with local and international NGOs and universities. The Climate Change Think Tank is the latest boost to this capacity.
>
> The development of the country's new constitution after Apartheid provided an opportunity to establish a modern bill of rights and responsibilities for all citizens of South Africa. As such it is a source of pride.
>
> Since its development, however, the world has moved on, with the divisions between national and municipal responsibilities becoming blurred. For instance the constitution and some of the subsequent legislation render city action on

> climate change complex. For example, partnerships between cities like Cape Town and companies on locally generated renewable energy requires support from all three spheres of South African government, a situation that impairs progress on achieving climate mitigation.
>
> *(Monaghan, 2011d)*

This dilemma remaining, Cartwright, citing De Visser (2005), points to the consequences of Cape Town's conservative interpretation of national public finance legislation for the type of decision required to tackle climate change. He also makes the case for more boldness by the city's leadership to challenge the national stance and adopt a more progressive interpretation on green procurement.

In concluding, Cartwright notes that:

> When it comes to city-scale decision making, the assumption that the future will be much like the past is no longer applicable for many African cities in the context of climate change. For Cape Town to continue to meet its local challenges and for it to become an example to the rest of the world, it is necessary for the City to keep updating its institutional capacity for sustainable development and climate-smart planning. Cape Town's Climate Change Think Tank is just the latest example of the City seeking to do this.

For further information visit http://africancentreforcities.net/ and http://www.capetown.gov.za (accessed 01 June 2011).

Contrast Cape Town's situation with the powers bestowed on peers elsewhere and one can see that the South African city is fighting with one arm tied behind its back. Surely it is in the national interest as well as local interests for Cape Town to be more self-reliant in this regard? Cape Town's ability to attract inward investment means central government grants can be directed to less able areas. Conversely, take, for instance, the state of California, USA, which has the legal power to demand suppliers provide renewable energy.

California establishes the USA's highest renewable power target

New legislation signed in 2011 means that electricity providers in the state of California must obtain 33 per cent of their power from renewable sources by 2020. It is the highest goal set in the country and is an upgrade of the previous law which set a 20 per cent target (Associated Press, 2011). The higher standard is intended to reassure investors about the certainty of California's long-term commitment to alternative energy sources.

So, in conclusion to the notion of a harmonised constitution, it is seen that two elements are essential for consideration. One is the extent to which local and national constitutional frameworks integrate in order to genuinely achieve the highest level of subsidiary to local government or the community. The other is in terms of the scope of aspects it must deal with (be it in respect of environmental limits or human rights) for tomorrow as well as today. Taken together then, what would such an enabling framework look like?

An enabling constitution for local leadership

To further tease out a solution to the conundrum of harmonious constitutions it is helpful to consider the findings from a global study comparing and contrasting 16 different countries' flagship climate legislation and/or plans including Brazil, China, France, India, South Africa and the UK (Townshend et al., 2011). Results show that adaptation (changing the way we live to respond to fluctuating weather as opposed to mitigating against those changes) was not a main focus of any, although it was mentioned to some degree in 75 per cent of them. Furthermore, the study goes on to consider the role of the lead parliamentary committee through which the legislation is processed, which it states is reflected by the primary motivation for the legislation, for example, in fossil-fuel-rich Russia the responsibility is combined with that for natural resource management, which may lead to a potential conflict of interest.

Learnings from this are three-fold. Firstly, there is a lack of long-term vision for future generations of society. Secondly, it shows there is a disconnect between national decision-making and local needs, given that councils appear not to have a strong influence over policy which will greatly affect them or indeed what they are most likely to be tasked with implementing. Thirdly, it also raises questions over the governance arrangements associated with the drafting of legislation, such as the influence of the private sector (a topic which will be further explored in Chapters 7 and 9).

It is such myopia that has led to a number of radical manifestos emerging around the world with the specific aim of updating national constitutions or law to ensure decision-makers recognise the plight of future generations (Garton Ash, 2011), that is, to ensure our children and grandchildren are not less prosperous, less free or less secure than today's citizens. Notable policy developments have been undertaken in Bolivia, Hungary and Kenya, and each is now discussed further:

- In Hungary, since 2008, the Parliamentary Commissioner for Future Generations ombudsman has been elected by parliament to protect the constitutionally guaranteed fundamental right to a healthy environment. The Commissioner receives petitions from those concerned that this right has been or may be violated, upon which he or she investigates, as appropriate, and makes recommendations for change to the public body concerned (Roderick, 2010). Hungary is considering building on this platform, by giving mothers additional votes in elections so they can vote on behalf of their children (to a maximum of one child) (Phillips, 2011).
- In Kenya, since 2010, environmental rights have been explicitly recognised as a part of the revised constitutional relationship between its government and citizens. This includes the right to a clean and healthy environment, and in particular the

right to have the environment protected for the benefit of present and future generations (Green Futures, 2011).

- Bolivia, in 2011, was preparing the world's first law granting the natural environment equal rights to humans, with the country's rich mineral deposits redefined as 'blessings'. This includes the right of nature not to be affected by large infrastructure projects that hurt the balance of ecosystems and the local inhabitant communities (Vidal, 2011a).

The success of these policy developments from Africa, Latin America and Eastern Europe has inspired similar calls for change about constitutional and parliamentary reform in the UK from a consortium of NGOs led by WWF and the Foundation for Democracy and Sustainable Development (Roderick, 2010). This is explored in more detail in Box 6.1.

Box 6.1 Environmental limits legislation

Living within environmental limits – the critical point at which pressure on a system creates irreversible change to the detriment of humans and others to which it provides a service – is a key aspect of sustainable development argues a new collaboration of UK environmental NGOs called the Alliance for Future Generations. So, as part of taking a longer view in terms of UK parliamentary governance options for a finite planet, one idea put forward by some participants of the Alliance, which includes WWF and the Foundation for Democracy and Sustainable Development, is the enactment of an Environmental Limits Act (Roderick, 2011).

Roderick proposes that such an Act would require the Secretary of State to monitor and report to Parliament on the country's effects on the Earth's system processes, both in absolute terms and also relative to other countries. He or she would be assisted in this task by a Planetary Boundaries Committee, which would have a wide-ranging advisory and research role, as well as a duty to make recommendations that the Secretary of State would be legally obliged to follow (or to provide reasons for not doing so).

The Act would identify 9 critical planetary boundaries or limits: climate change, biodiversity loss, biochemical cycles (e.g. nitrogen, phosphorus), ocean acidification, water consumption, land use, ozone depletion, atmospheric particulate pollution and chemical pollution (Lynas, 2011).

Critically, it is argued here by Roderick (2011), this approach has a precedent in the UK with the Climate Change Act 2008.This Act took the politics out of the science by having an independent body that sets national legally binding 'carbon budgets' and publically reports on the government's progress against these targets.

However, as identified earlier, national interventions need also to be translated to the local level, but how would this occur? One route may be that which is being adopted by Totnes in the UK.

Transition town of Totnes (UK)

Totnes is a town in south-west England. It launched the Transition Town Totnes (TTT) initiative in 2005, with the goal of providing a blueprint to enable communities to make the change from an oil-dependent life in order to become more self-sustaining (Siegle and Borden, 2011).

The transition idea works on the basis that waiting for government to act on peak oil (the point at which demand exceeds supply) will be too late, and that acts by individuals will be too little. So working together as communities is where the real change will happen; that is, transition is bottom up, a culture change that is about people coming together to do the work for themselves.

Central to the whole plan (based on scenarios to 2030), are ideas including permaculture gardening scaled up to bring food resilience to town centres and a local currency to prevent money leaking from the local economy and ensure more wealth is retained.

Whilst such big and courageous local initiatives (such as this in Totnes) should be welcomed, caution that any devolved power to citizens and communities must still be accountable must be taken into account; that is, the role of democratic and accountable local councils is key, given one cannot vote against a resident-led social enterprise or a charity (McInroy, 2011). This recognition of the importance of subsidiarity could be recognised through the transfer of budget control from central to local government, including localised tax-raising powers (Conrad, 2011) and law and order powers, which would be more efficient, less duplicative and less fragmented.

This does not mean, however, that local council executives are all-powerful and that residents cannot have a greater say in plans for city mayors or how council budgets should be spent (Wilson, 2010; Wintour, 2010). Further, voting processes for electing local councillors may hinder long-term decision-making. An example of this is that in the UK some local authorities have the option of electing only a third of their councillors ever year as opposed to all councillors every 4 years (DirectGov, 2011), which may only result in a permanent cycle of myopic electioneering instead of big and bold forward planning. After all, only long-term planning will deliver the needed transition to a low-carbon society.

When considering new rights and responsibilities afforded to citizens, it is also important to respect the principle of subsidarity so that strong local government is not undermined. One way of doing this is through 'co-production'.

Co-production is about making good use of all the resources in society, not just those in the public sector, in order to improve the quality of life of all citizens by changing the relationship between local government, residents, service users and community groups. It means hardwiring into a public service – such as health, community safety or crime prevention – the know-how and loyalty of local users in order to deliver more meaningful outcomes (e.g. neighbourhood watch schemes). To smooth the transition, however, joint governance is paramount too, through a process of co-design, co-commissioning, co-delivery and co-assessment (Governance

International, 2011); that is, it is an enabling mechanism, whilst also having 'teeth' to monitor and remedy any errors or misdemeanours in provision of a valued public service.

Such co-production in certain manifestations is increasingly de rigueur; as discussed in Chapter 2, PPPs have grown eightfold over the past decade or so, and are something which could be accelerated again over the next decade given the calls in many quarters for big business to help the public sector lead the drive to a green economy and sustainable urban development. But, if this holds true, what are the upsides and downsides to such large scale co-production initiatives? What is the interconnection here between wider urban development strategies, such as a move to densification? These questions are considered in Chapter 7, as part of the movement for 'just cities'.

Conclusions

- Rights *and* responsibilities make up a constitution, but rights are often overlooked.
- A clear understanding of peoples' responsibilities as well as rights will deliver a 'we' rather than a 'me' culture.
- To ensure local accountability, the principle of subsidiarity (democratically elected government at the lowest possible level) must be observed.
- But to enable this to work there needs to be harmony between the national and local constitutions. Such enabling constitutions would consider the needs of future generations.
- Co-production of services by community members for community members is an ideal route to ensuring quality provision that meets everyone's needs.

Part Three
Just cities

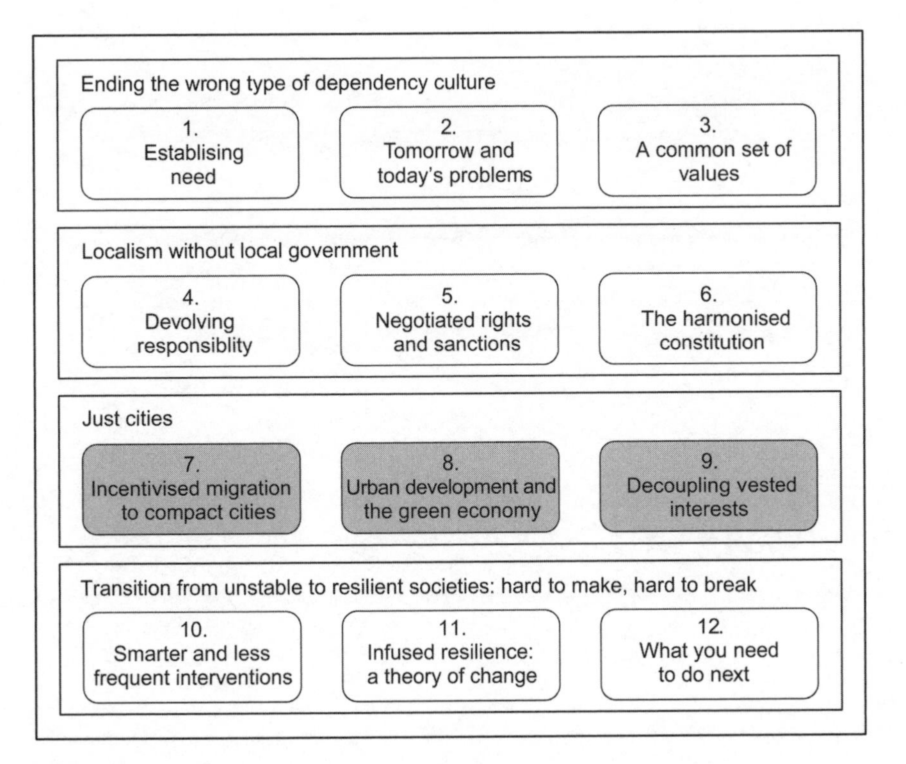

7

INCENTIVISED MIGRATION TO COMPACT CITIES

> It is about bringing the eco city directly to the community.
>
> Mi Kyung Moon, Environmental Policy Adviser, Changwon City (Moon, 2011)

Reaffirming the need for compactness

Although cities occupy only 2 per cent of the world's crust, 53 per cent of the global population resides in them. Cities also account for 80 per cent of global GDP, and 600 of these cities are home to 20 per cent of the world's population, generating 60 per cent of global GDP (Dobbs et al., 2011). At the same time, 33 per cent of all city dwellers live in slums and cities are also a vast consumer of resources, producing 75 per cent of world CO_2 emissions. So, cities rule yes, but some cities are more equal than others and which ones rule is set to change.

By 2025, according to Dobbs et al., the membership of the group of the top 600 cities will change as the momentum for migration moves from the developed nations to the south and east and overwhelmingly from China (e.g. Guiyang) as well as India (e.g. Surat) and Latin America (e.g. Cancún). This change is also anticipated to result in a further concentration of growth with a new top 100 cities contributing to approximately 33 per cent of GDP growth. More than this, though, the trend for 'midsized' cities (150,000 to 10 million inhabitants), as opposed to 'mega' cities (more than 10 million inhabitants), to deliver most growth will continue and increase to the point they will deliver 40 per cent of global growth by 2025 (Dobbs et al., 2011).

To put the power of cities in context, it is interesting to compare their economic might to that of nations and big business, as detailed in Table 7.1, an extract from the world's top 100 economies of 2008.

TABLE 7.1 Economic power of cities

Ranking	*Country/City/Company*	*Country/City/Company*	*GDP/Revenues (US$bn)*
1	USA	Country	14,204
2	China	Country	7,903
3	Japan	Country	4,354
7	UK	Country	2,176
12	Tokyo	City	1,479
14	New York	City	1,406
32	Royal Dutch Shell	Company	458
67	Toyota Motor	Company	263
90	General Electric	Company	183
100	Barcelona	City	140

(Source: Adapted from World Bank, 2010a).

As is evident from this extract, cities matter because they are large economies in their own right (World Bank, 2010a) and, therefore, have a huge impact on prosperity. For example, the city of Tokyo accounts for half of all of Japan's national economic power and is nearly 6 times larger than its biggest company, Toyota Motor. The power of cities is not surprising given that, according to Glaeser (2011), humankind's greatest asset is our ability to learn from the people around us and so cities are a catalyst for innovation and entrepreneurism, connecting as they do employers with employees, and entrepreneurs with customers.

Naturally, companies and developers looking to invest will seek out cities that offer the best opportunities for growth (such as Tokyo, New York and Barcelona), which means their importance grows yet further. But, from a socially and environmentally progressive stance, given the great migration to cities that is already happening, should exponential city growth be encouraged and what *type* of city growth strategy do we want to see?

There is an emerging body of compelling evidence that city dwellers are greener, enjoy greater social mobility and make it easier for local authorities to run cost efficient public services than their rural cousins. The average Londoner produces around half the emissions of the average Briton, the average New Yorker produces 30 per cent of the emissions of the average American and people in Sao Paolo produce just 18 per cent of the average Brazilian. Women in cities on average get better jobs as they stay in school longer, have better access to contraceptives, get married later and have their first children later (Barley, 2010). Installing a sewerage network along a city street, which can connect thousands in one go, is much cheaper than connecting rural conurbation to rural conurbation if they are far away from each other.

Furthermore, average incomes in countries that are more than 50 per cent urbanised are at least 5 times higher than those in countries that are less than 50 per cent urbanised. Infant mortality rates are 3 times higher in the less-urbanised countries. The metropolitan wage premium is 3 times larger for workers with more than 15 years of labour market experience than for workers with less than 5 years experience (Glaeser, 2011).

In terms of resilience against fuel security needs, take for instance the example of New Orleans.

Saving residents money as gas prices rise in New Orleans (USA)

In the face of rising gas prices in 2010/11, research released by CEOs for Cities (2011) concluded that compact cities are best placed to save residents money. Compact development patterns and extensive transport systems enable their residents to drive fewer miles, on average, than the typical American. Examining 10 urbanised areas, New Orleans came out on top in the study.

Accordingly, is it sufficient to simply say that we should therefore incentivise or even enforce an urban life over a rural one? Some say yes. Take for instance the example of Ordis in north-central China.

Paying farmers to move to Ordis City (China)

To try and halt the loss of habitat to the desert in water-scarce north-central China, the government has paid farmers and shepherds to move to the district capital, Ordis City. Approximately half of the region's 435,000 population have migrated as a result. The rationale is that people will have a lower environmental impact by living in a dense city rather than spread across rural areas. The city's new inhabitants also benefit economically – they can earn up to 5 times more money in the city (Barley, 2010).

Of course, practice may vary from city to city, even within the same country. Contrast for instance the approach in Ordis to that of Beijing, which has reportedly implemented a new set of socially regressive measures to calm property prices and traffic congestion that includes preventing migrants to the city, who account for 33 per cent of the city's population, from buying homes and cars (Economist, 2011).

So does this then mean that *any* kind of city migration should be welcomed or are certain types more desirable? In short, the answer is the second option – migration to *compact* cities.

Through smart densification, such as compact cities, as well as being better positioned to realise the benefits outlined previously, it is also possible to avoid the considerable negative effects of urban sprawl. In terms of the benefits, so-called 'economies of agglomeration' occur whereby wages, productivity and innovation rise with densification. For instance, there seem to be more labour competition, knowledge spill-over and less congestion (Roberts, 2011).

In terms of the latter, greater concentration means more metropolitan areas are more accessible on foot, which, coupled with effective transit systems, increases economic justice by widening the job market for low-income people without cars and also helps to tackle growing obesity and diabetes problems and reduce the high number of injuries suffered as a result of automobile accidents (Goodyear, 2010).

Box 7.1 and the following resilience learning from Lagos look at traffic deaths and transport supporting access to the job market respectively.

Box 7.1 More children killed by traffic than by diseases

Globally, road accidents take the lives of 3,500 people every day, 3,000 of whom are from developing countries. The FIA Foundation (Ramesh, 2010c) predict the death toll will increase to 5,700 a day in a decade.

More lives among those aged 5–14 were lost on the roads than to malaria, diarrhoea and HIV/Aids combined, yet unlike these diseases traffic deaths are mostly absent from the development agenda, say the report's authors, and as such undermine MDG goals such as universal primary education.

Possible interventions to address these shocking statistics include separation of traffic streams from pedestrians, building of raised kerbs, introduction of speed reductions in urban areas, and laws requiring motorcycle riders and passengers to wear helmets.

Changing the lives of millions through the railway in Lagos (Nigeria)

Rural–urban migration will see Lagos overtake Cairo as Africa's biggest city in the next 5 years with a population of 12.4 million. Yet, Lagos is badly in need of mass rapid public transport. It is not unusual to find Lagosians walking for over four hours to get to work.

It is hoped that a rail renaissance through EkoRail can be part of the solution to this difficult way of life. The new lines will carry 1.4 million passengers per day. Crucially it will be cheaper, faster, safer and more reliable than the current alternatives (Smith, 2011).

Whilst Lagos aims to follow in the footsteps of the Latin American cities of Bogota and Curitiba (Monaghan, 2010) on the successful delivery of mass rapid transport systems, concerns have also been raised as to how much of the money being generated through such enterprise is kept local (Smith, 2011). For instance, as this development in Lagos is funded and led by Chinese investors and operators, how can Nigeria balance this important access to new infrastructure (and associated investment) with the need to boost their own employment figures and the skills and competencies of its own workforce? The issue of keeping money local is explored further in Chapter 10.

For all this to work, of course, the stereotypical downsides of cities compared to the countryside – that they are soulless, depressing, dangerous and dirty – must be addressed. The provision of high-quality leisure facilities, public amenities and anti-grime and anti-crime measures needs to be a key priority. Amman is one city challenged with dealing with some of these dilemmas as part of a constant influx of migrants.

Refugees and the pressure of rapid urbanisation in Amman (Jordan)

The capital, Amman, is a city that has been pressured into transforming towards a more sustainable form of master planning driven by the power of urbanisation and diminishing natural resources (UN-Habitat, 2011). The country is 78 per cent urbanised, with the majority of the inhabitants aggregating in Amman. Its location in the Middle East means it attracts the highest number of migrants in the world, with over 60 per cent of the Jordanian population classified as migrants.

Added to this, Jordan is one of the top four most water-scarce countries in the world, with quality and quantity both representing a dilemma; along with other problems including desertification, soil degradation and deforestation, rising costs of electricity and other fuel. Traditionally, the planning of Amman has been focused around the use and dominance of the car, resulting in significant CO_2 emissions.

Consequently, resource use is an increasingly important economic concern for the government and it is now looking to advance a number of solutions:

- A transport and mobility master plan has been developed.
- Use of solar energy sources is being investigated.
- Grey water is being used for irrigation.
- A waste-minimisation approach has been implemented, including deriving energy from waste.

Subsequently, the Amman Green Growth Programme was established as a city-wide plan of action to aggregate and reduce emissions across water, waste, public utilities, transport and forestry sectors. Key aims are to improve the urban environment, adapt to climate change and increase the cost-effectiveness of municipal services, as well as revenue through carbon markets.

So, how are the leading cities attempting to devise the smartest compact policies?

Ensuring smart density wins

In recognising the need to decouple development from natural resources use, smart density policies must use planning approvals to regulate by redirecting land use into more productive capacity (Newman et al., 2009). Further to this, the London School of Economics (LSE) has identified a number of interventions to improve the efficiency of intra-urban resource flows and improve prosperity of life (UN-Habitat, 2011). These include:

- establishing urban growth boundaries to limit urban sprawl;
- land-use regulations that promote redevelopment of city areas and protect green space corridors;
- density regulations to enforce minimum densities;

- density bonuses for developments that support city-wide sustainability;
- special planning powers for urban development corporations or urban regeneration companies;
- vehicle and traffic regulations to reduce emissions, use of fossil fuels and congestion;
- maximum parking standards to discourage private car use;
- incentives for car-free developments;
- minimum energy efficiency and emissions standards for buildings and vehicles.

Within this, the LSE (UN-Habitat, 2011) also recognises that in cities with large informal settlements – ranging from 25 per cent of the urban population in Asia and Latin America to 60 per cent in Africa – special ways will have to be found to establish infrastructure services (e.g. water, sanitation, energy) that remain affordable for poor households and informal commerce.

Bollier (1998) also highlights one of the most influential forces propelling sprawl – the tax system – whereby local government's reliance on property taxes as a chief source of revenue means they can deliberately, or even unintentionally, adopt non-compact city policies. Here, through fiscal zoning (tax bands on properties), cities prefer to develop expensive homes and commercial industrial properties regardless of whether they sit vacant or underutilised and so only encourage sprawl.

In reality, however, the business rationale to act, the governance levers available to city leaders and the particular migration context will vary from country to country. An interesting comparative analysis then is an applied research project on compact city policies led by the OECD, where one such city was Toyama, Japan, which is facing the challenges of how to ensure a depopulating society remains compact (Monaghan, 2011e).

The challenges of compact policies in Toyama (Japan)

Toyama is located in central Japan, with the nearby Tateyama mountain range to the east and the Toyama Bay to the north. A mid-sized city with approximately 417,000 inhabitants, Toyama's population is shrinking and there is an over-reliance on the car due to urban sprawl and the public transport system in decline (something which is likely to inconvenience its citizens further as they continue to age and may become unable to use a car freely). These factors have necessitated a renewed appetite for compact city planning (OECD, 2012a).

Consequently, to support intercity knowledge transfer, Toyama has been participating in an OECD-led project since 2010, along with Melbourne (Australia), Paris (France), Portland (USA) and Vancouver (Canada), to compare the impacts and challenges of compact city policies to support 'green growth' (Matsumoto, 2010).

The premise to the applied research, which will report its findings in 2012, is that a compact city is an approach to an urban form that can make better use of resources at the metropolitan region scale through:

- dense and contiguous development patterns;
- built-up areas with mass-transit linkages;
- accessibility to local services and jobs.

According to the OECD (2012b) this in turn enables green growth by:

- shortening intra-urban travel distances and thereby reducing CO_2 emissions and improving productivity;
- reducing automobile dependency and encouraging citizens to use public transport systems or active commuting;
- consuming less energy (outside of transport energy consumption) with promoting district-wide energy utilisation and local energy generation;
- increasing the efficiency of public service delivery;
- allowing better access to diverse local services and jobs, which leads to a higher quality-of-life and creates more favourable conditions for growth.

Tadashi Matsumoto, Senior Policy Analyst, Regional Policies for Sustainable Development Division, OECD, says:

> Cost-efficient public service and better mobility are the prominent components of Toyama's business case in seeking to secure local support for its compact city policy. The shrinking population in Toyama also means a shrinking tax base. High car dependency in a dispersed city for an ageing population less able to drive means more walkability is a must.
>
> *(Monaghan, 2011e)*

Despite this clear need, key challenges for cities like Toyama remain.

> Sometimes a compact city policy is too decentralised, so there can be a need for the metropolitan and even the central government to take a stronger role. Conflict management over land use, particularly in relation to green spaces and new developments in the urban fringe, needs to be more effective. But most of all, packaging of policies needs to be improved so they are multi-dimensional and complementary, to avoid single purpose policies – for instance just protecting green space or retaining farming areas – exacerbating the negative aspects of compact policies, such as traffic congestion and housing affordability, and to prevent economic measures being counter-productive – for instance congestion charges without mass-transit improvement contributing to urban sprawl ...
>
> Emerging data from cities with mature compact city policies like Portland and Vancouver suggests that strategic policy-making in this way does work, most notably strict controls on new development with mass transit improvement, which has led to an increase in built-up areas with lower CO_2 emission and increased productivity. More applied research is needed, however, to link this particular intervention with this outcome, given it may in part be due to

> other actions too. Moreover, the effectiveness of such a particular intervention would be different according to the local circumstance. This is key focus for the next phase of OECD's work.
>
> *(Monaghan, 2011e)*

For further information visit http://www.oecd.org (accessed 28 June 2011).

What is particularly insightful about the case of Toyama in comparison to other compact cities is that it reminds us, firstly, of the need to overcome discipline silos (discussed further in Chapter 11) and, secondly, the need to pursue harmony between national and local governance systems (as considered in Chapter 6). So, whilst compact cities may be the right direction of policy travel, they are not always desirable to municipal leaders, as witnessed in Delhi.

Green does not always mean good in Delhi (India)

In recent years there has been a huge drive towards making Delhi a 'world class' city, one element of which is a 'clean, green' campaign. Whilst good environmental progress may have been made in terms of air pollution through low-carbon public transport, some of the process involved has been socially regressive, argues the University of Cambridge (2011) amongst others.

Slums and informal settlements have been demolished even when they were legal dwellings, to be replaced by supposedly 'cleaner and greener' malls and elite housing apartments (which, ironically, have been constructed illegally). As a result, poorer people have become even more marginalised as they lose their livelihoods as the opportunities for small-scale manufacturing and street vending are driven out of the city.

Thus, whilst humans may be the most adaptable of species, and have generally embarked willingly on this great migration to cities (many of these migrants are not passive victims, but rather opportunists taking a calculated, and often educated, gamble on an urbanising world; Saunders, 2010), it is also imperative that social justice is observed as cities grow and migration continues. City and countryside dwellers have inalienable rights and responsibilities which should be enshrined in any constitution developed (as proposed in Chapter 6) to protect slum dwellers and penthouse residents alike.

To be deliverable in the long term, however, compact cities will need to be developed by multilateral partners – one way will be through the private sector and the development of the green economy.

Conclusions

- Cities currently dominate, both economically and in terms of population. Vast migration to cities is already well under way.

- Any migration from dispersed rural areas to dense cities is potentially good for the environment and social mobility and so should be encouraged.
- To be effective, however, compact city policies that are able to deal with the complexities must be in operation. This means understanding the connections between an array of issues from developer tax contributions and urban sprawl through to how equality of access can build social cohesion.
- Urban leaders must ensure the city secures a good deal from these developments and that all parts of society benefit.

8

URBAN DEVELOPMENT AND THE GREEN ECONOMY

> There was no conversation 100 years ago about the 'high' carbon economy. Whilst the next wave of urbanisation must be sensitive to the transition to a 'low' carbon economy we also need to talk about it in a way that meets wider social and developmental aspirations too.
>
> Dr Mike Hodson, Research Fellow, Centre for Sustainable Urban and Regional Futures (commenting on the need to ensure that the objective of a green economy is relevant and desirable to ordinary people) (Monaghan, 2011f)

In Spring 2011 the near financial collapse of Southern Cross, a major private social care provider in the UK which looked after 31,000 elderly residents in 750 homes, sent shockwaves through the country's political and market establishment. As with the banks, so with care homes: a privately operated but socially vital activity was mismanaged and, whilst the shareholders gained in the good times, the taxpayer could, again, be asked to cover losses during the bad (O'Grady, 2011).

So, if it can happen in financial services and social care, is there the same danger for private sector involvement in major urban infrastructure projects – such as rapid surface transport, smart grids and renewable energy or water sanitation schemes – as part of the green economy? In short, yes.

This is an important question because this is a crucial area for growth – according to the US Pew Environment Group, 2010 was a landmark year for global clean technology investment, reaching a record US$246 billion and this was the first time that investment in renewable energy overtook nuclear power (Harvey, 2011). The race for new money and jobs in the so-called 'green economy' had come of age, as evidence by China's latest five-year plan in Box 8.1.

Box 8.1 China's five-year plan 2011–15: a switch to green growth?

China's 2011 economic blueprint outlined major initiatives to green the economy over the coming half decade (Watts, 2011). This was interesting as it shared commitments including:

- energy efficiency and environmental services being declared 'priority industries' for the first time;
- a doubling of the amount being spent on environmental protection to US $0.5 trillion;
- a carbon-intensity target being set, measuring the ratio of GHG emissions relative to GDP;
- a new environmental tax on heavy polluters to levy fees on discharges of sulphur dioxide (SO_2);
- a mandatory carbon trading system on a regional level.

At the same time, another major step is the gradual shift away from GDP-based performance evaluation (due to threats to China's water supplies and food security from environmental damage arising from unsustainable development) (Watts, 2010).

The timing of this dramatic growth in clean technology markets was even more interesting, given that the major urban centres that would be the focus of many of these clean technologies were simultaneously facing savage public spending cuts, and so were looking to the private sector to fill this gap.

At the same time, there were rumblings from the UN that they would abandon any hope that a single deal on climate change could soon be reached, and would instead switch to proposals for a green economy on the basis that this was much more politically palatable (ENDS, 2010; Goldenberg, 2011). As of 2011, the continued absence of a meaningful climate deal is unsurprising if we consider the narrative on national versus per capita emissions between China and the USA (as set out in Box 8.2).

Box 8.2 World carbon emissions

Human dependency on fossil fuels has led to excess amounts of greenhouse gases such as CO_2 being released into the atmosphere, trapping heat and resulting in the warming of our planet. This is bad news for everyone, especially for those in poverty (given they tend to live in developing countries that will face the brunt of the chaos and are less well equipped to cope). As the charts show, emissions vary wildly per capita and per country.

(Adapted from IEA, 2009)

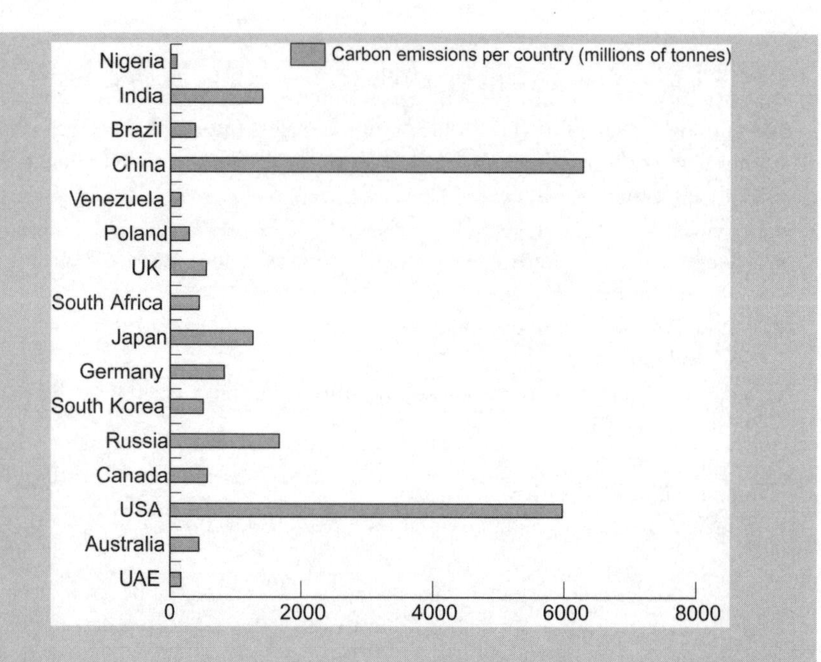

As these charts show, whilst China as a *whole* produces more emissions than the USA, emissions arising from each individual Chinese citizen are just 20 per cent of those of citizens in the USA. At the same time, a record rise in emissions (despite the recession) means the goal of preventing a temperature rise of more than 2C – which scientists say is the threshold for irreversible

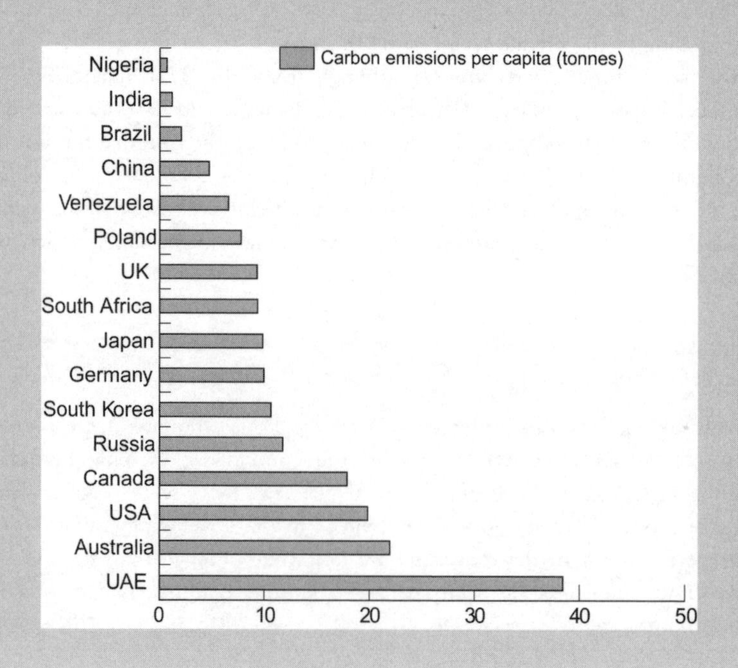

climate change – is most likely beyond us, says the International Energy Agency (IEA, 2011). This raises challenges in terms of sustainable consumption and fairness, as, for instance, if China continues to grow apace then increased standards of living may lead to more consumption and, thus, more emissions. But is it fair for US citizens to expect their Chinese counterparts not to aspire to their standard of living? If not, perhaps a more desirable outcome for all is development that does that does not involve killing the planet (Monaghan, 2010).

Of course it is by no means unique for a call to arms to be made to business by politicians.

The USA called upon the private sector twice in the last century to kick-start the economy – first in the Great Depression in the 1920s and again in the 1940s when it turned to the business community with a request to assist the war effort (e.g. Ford were asked to move to producing tanks instead of cars). In the 1960s, European religious leaders and NGOs called for companies to halt investment in South Africa due to their Apartheid policy. In the 1980s, when product eco-labelling began to be mainstreamed in North America and Europe, as a response to unsustainable consumption and irresponsible marketing companies were pressurised to act by NGOs and the media campaigns (e.g. on canned tuna, timber furniture). Then, in the 1990s social auditing emerged as a new panacea to hold the power of multinational enterprises to account (Monaghan et al., 2003; Brown, 2011). And, so, the world's problems are again being laid at the feet of big business.

Yet, despite the grave responsibility being placed on the interface between the green economy and urban development by the UN High-level Panel on Global Sustainability (2011) and others, there remains a surprising lack of clarity about how the green economy is being defined, what role (if any) is being asked of local government and how any new forms of public–private partnerships (PPPs) are to be overseen.

The recognition by national governments of the important role of local government in any transition to a green economy varies from country to country. To illustrate this point further, in Box 8.3 the approaches of the UK and the USA are contrasted.

Box 8.3 Enabling the transition: expectations of local government?

In 2011, the UK set out an ambitious green economy roadmap to the year 2050 (HM Government, 2011). One of several important statements in the document is a commitment to develop a 'green policy framework' so that policy making about the green economy across a number of departments – transport, energy, business and environment – is clear and effective.

The strategy also stipulates what the government expects the contribution from the UK business community to be in return, which is very laudable. Interestingly, though, the government has not yet said what role, if any, it expects local councils to play in enabling this transition, be it in relation to local planning controls, procurement or workforce skills.

By way of comparison, it is interesting to note how the US green jobs market has succeeded by adopting a different approach (McNeil and Thomas, 2011), which involves local government playing a crucial enabling role in the transition to a greener economy. For example, to help stimulate the energy-efficient lighting industry, the state of California has collaborated with utility companies, trade unions and electrical contractors association to develop 'journey upgrade' training for installers. Targeting training to electricians (who must be licensed in California) and limiting participation in incentive programmes to those who have achieved a standard of training has enabled contractors in the advanced light controls industry to compete on the basis of quality for new construction and retrofit developments. McNeil and Thomas (2011) conclude that the US experience shows that real gains can be made through such strategic, local, bottom-up interventions that combine strong policy knowledge with grassroots endeavour.

In terms of oversight of new forms of PPPs, this relates to a clear grasp about the wonderful opportunities and downside risks of these novel modes of development. This should include primarily who the likely winners or losers would be from this transition so changes can be sensitively managed (as illustrated in the swot analysis in Table 8.1). For instance, whilst there are tremendous benefits for city leaders in securing much valued finance and technical expertise from the private sector, it is also crucial to ensure that any increase in prosperity benefits the poor as well as the affluent and, in particular, helps the Southern poor too, otherwise any single grand deal put together by the UN on the green economy would go the same way as that on climate change – most likely, nowhere at all.

What this means, in simple terms, is that, whilst city leaders must be brave and steel themselves for taking on the big social challenges of the day, by embracing the green economy, they must also be cautious as to which inhibitions they choose to shed. Despite the focus, it is not clear in practice what they need to do, for instance with regard to the protection of domiciled workers. Trade unions have highlighted their concerns on this issue, arguing that certain regions are being cited as 'best practice' green economies though all labour rights are not recognised (such as in Korea) (Council of Global Unions, 2009), and extolling exemplars they consider to be pro-labour, such as in Brazil (Keivani et al., 2010). Another is the USA, explored in more detail using a case from Pennsylvania which is creating employment despite the economic downturn.

Green works for the economy in Pennsylvania (USA)

Pennsylvania is rated by the Pew Charitable Trust as amongst the top three states in the nation that have capitalised on revitalising their economy through decarbonisation (Dean, 2011). This rating was attained through projects that:

- provide alternatives to carbon-based energy sources;
- conserve the use of energy and all natural resources;
- reduce pollution (including GHG emissions);
- reuse waste.

TABLE 8.1 The interface between the green economy and urban development: so far, so good?

Strengths	*Weaknesses*
High impact economic transformation – new inward investment in major 'sun rise' technologies (e.g. decarbonised rapid surface transport).	**No universal definition** – confusion between 'green' and well-adapted economies.
Broad popularity – new jobs, skills and infrastructure (no global climate deal is imminent anyhow).	**Wealth versus prosperity** – the winners and losers from the new 'gold rush' to the green economy are unclear (e.g. bio-fuels exacerbating food scarcity for the poor).
Complementary to densification need – economies of scale for all, ensuring wider benefits (e.g. equality, resilience).	**Inappropriate deals** – concerns that poor and/or inflexible multi-century deals may be negotiated with industry due to spending cuts and economic recession (i.e. pressure on cities in the USA predicted to 'go bust').
Opportunities	*Threats*
New finance in an age of austerity – new capital/revenue brought in but also managed more effectively to save or make more money (e.g. special purpose vehicles for leasing, debt, outsourcing with 8-fold growth in PPPs over past decade).	**New source of NorthvSouth trade disputes** – no clear agreement on what 'green economy' means or how it will aid development (e.g. Will it be decarbonised? Will it be fair?).
Dramatic skills and knowledge transfer – access to best in class ranging from Cisco to IBM (e.g. shared distribution).	**Unaccountable companies running cities** – plethora of macro- or micro-public–private partnerships and 'CSR-urbanisation' codes led by business. Focus remains on short-term profit and governance arrangements remain weak.

© Monaghan, 2012

A key trait of the Pennsylvanian approach is the proactive effort to build collaborations between the private sector, universities, not-for-profits and community organisations. While the state is spending funds and receiving federal dollars, they do so when those dollars are matched, if not exceeded, by private investment, or where the state can see measurable results in job growth and revenue.

One example of these collaborations is GreenWorks, a for-profit company that works to enhance the quality of life in the region by creating new communities in previously developed urban areas. Since 2005, they have developed over $US25 million worth of projects in downtown Harrisburg and Carlisle, two of the many US cities that got left behind in the rush to develop suburbs and malls. Their projects provide renewed business space, occupied by emerging and small businesses. They actively seek out brownfield sites – that is, land that has been abandoned and where redevelopment may be complicated by environmental contamination – remediating old problems with new ideas. Using solar, geothermal and wind, they renovate old or build new energy-efficient certified buildings. By focusing on the surrounding community, GreenWorks integrates community development, reinvigorating 'walkable' live, work, play and learn neighbourhoods.

As a result of the drive to transition to a green economy, Pennsylvania was one of a handful of states in the USA that managed to grow employment in 2011 by over 1.1 per cent.

However, although Green Works is one example of how the 'green economy' can work in practice, further challenges are raised through its very definition.

Problem of definition

There have been varying approaches to actually deriving a common understanding for the term 'green economy'. The United Nations Environment Programme (UNEP, 2011b) defines it as 'one that results in improved human well-being and social equity, while significantly reducing environmental risks and ecological scarcities'. The international think tank the World Future Council (Göpel, 2011) has adopted a different tack by working with a coalition of NGOs to draft a set of 8 'guiding principles for the green economy' ranging from the 'planetary boundaries principle' through to the 'beyond-GDP principle'. Whilst both developments are welcome, well intentioned and well researched, their focus somewhat lags behind the fast developing and complex state of the green economy sector and, in particular, the rise of new soft and hard forms of public–private collaborations.

Greater clarity and consensus on what is meant by the green economy and how we want to harness the private sector's unique abilities is especially important as the volume and types of PPPs is evolving apace. Regarding the complex state of the green economy, as noted earlier in Chapter 2, one study estimates that about 50 PPPs were operating in the 1980s, which had grown eightfold to at least 400 by 2006 (McKinsey and Company, 2009). In respect of the latter, for instance, there has also been a proliferation in more novel or softer forms of partnerships between companies and urban areas, particularly through so-called corporate social responsibility (CSR) programmes, a number of which are described in Box 8.4, before being critically analysed in Table 8.2.

Box 8.4 The new 'gold rush': corporate social responsibility and urbanism

WBCSD's Urban Infrastructure Initiative

The World Business Council for Sustainable Development (WBCSD) is a corporate membership network that works to advance sustainable solutions across the globe. The WBCSD launched the Urban Infrastructure Initiative (UII) to help companies get involved at the beginning of the strategy discussion to explore how to solve the varied sustainable urbanisation challenges such as competitiveness, quality of life and the environment. For further details refer to http://www.wbcsd.org (accessed 18 June 2011).

Forum for the Future's Sustainable Cities Index

Forum for the Future is a leading sustainable development NGO in the UK, working with government, business and public service providers to lead the way to a better world. Forum's Sustainable Cities Index ranks the UK's largest cities annually on three broad baskets of environmental performance, quality of life and future proofing. The 2010 index report was funded by General Electric, who co-authored the foreword and contributed an essay. For further details refer to http://www.forumforthefuture.org/projects/sustainable-cities10 (accessed 18 June 2011).

IBM Smarter Cities Challenge

The Smarter Cities Challenge is a competitive grant programme by IBM awarding US$50 million worth of technology and services (over 3 years) to 100 cities with a view to them becoming more vibrant places to live. It is the largest single philanthropic programme at IBM and these grants are designed to address the wide range of financial and infrastructure challenges facing cities today. The Smarter Cities Challenge is part of a wider IBM Smarter Cities Initiative that provides professional services to the public sector around the world. For further details refer to http://www.ibm.com/smarterplanet/us/en/smarter_cities/nextsteps/index.html (accessed 18 June 2011).

Home Depot Foundation's Sustainable Cities Institute

Home Depot Foundation, the philanthropic arm of the US retailer, invests in the Sustainable Cities Institute (SCI) as part of its attempt to support efforts in environmental stewardship initiatives and provide proactive solutions to helping create healthy 'green' communities. One of SCI's guiding principles is that a sustainable community effort consists of a long-term, integrated and systematic approach to developing and achieving a healthy community by jointly addressing economic, environmental and social issues. The intention is that SCI will be a tool that cities can utilise for a holistic, long-term approach to sustainability planning and implementation of healthy communities. The SCI's advisory board includes representation from cities, as well as from the private and NGO sectors. For further details refer to http://www.sustainablecitiesinstitute.org (accessed 18 June 2011).

Cisco's Connected Urban Development

Connected Urban Development (CUD) was born from Cisco's commitment to the charitable Clinton Global Initiative to participate in helping to reduce global carbon emissions. The CUD program draws on expertise from the Cisco Internet Business Solutions group – its global strategic consulting arm – to demonstrate how to reduce carbon emissions by introducing fundamental improvements in the efficiency of urban infrastructure using information and communications technology. For further details refer to http://www.cisco.com/web/about/ac79/ps/cud/about.html (accessed 18 June 2011).

The Philips Center for Health and Well-being

The Philips Center for Health and Well-being is a knowledge-sharing forum that provides a focal point to raise the level of discussion on what matters most to communities and to citizens. It brings together experts for dialogue and debate aimed at overcoming barriers and identifying possible solutions for meaningful change that can improve people's overall health and well-being. A key topic of its focus is 'liveable cities'. Experts are a mix of Philips's own management team, academia and multilateral agencies such as the World Bank. For further details refer to http://www.philips-thecenter.org/livable-cities/ (accessed 18 June 2011).

By comparing and contrasting we can see that, whilst each may tend to have a common point of departure, in this case philanthropy, their primary end goals differ somewhat, most notably between access to new commercial markets or to help transition to a plant operating within its environmental limits. Of course, the two intentions sometimes coexist.

Most striking, however, is that, whilst city leaders may have some role in shaping these schemes, none of them are led by democratically elected city councillors or mayors (as shown in Table 8.2), thus potentially undermining the key principle of subsidiarity we referred to earlier. Whilst this does not mean that corporate-led or inspired initiatives such as these should not be welcomed – they should be applauded as meaningful contributions – urban leaders and policy makers need to be very clear about which ones they choose to support or engage with, and ensure that the power of these initiatives is kept in check with local democratic institutions.

Given this conundrum, what is the likely future viability of the green economy, and how can this dilemma be traversed?

A decarbonised economy the North and South can believe in

If PPP- or CSR-urbanisation schemes as seen previously continue to emerge and increasingly dominate the landscape, it would appear that a number of scenarios could ensue that determine the fate of the green economy.

Having seen how the green economy definitions, applications and governance are in question, perhaps the amber scenario shown in Figure 8.1 best depicts our current fractious path over the next decade. Here there will be benefits, but these will prove

TABLE 8.2 Strategic choice dilemmas for urban leaders: who to partner?

CSR Scheme / Features	*Governance*			*Strategy*			*Obligations*			*Policy*			*Impact*		
	Business-led	*City-led*	*Multi-stakeholder led*	*Access to new markets*	*Environmental protection*	*Green economy*	*Voluntary*	*Mandatory*	*Wider legal implications*	*Broad theoretical policy*	*Formal management guidelines*	*Issue specificity*	*Principle based*	*Process driven*	*Desired outcomes*
WBCSD: Urban Infrastructure Initiative	×					×	×				×			×	
FFTF: Sustainable Cities Index (with GE)			×		×		×					×			×
IBM: Smarter Cities Challenge	×			×					×	×			×		
Home Depot Foundation: Sustainable Cities Initiative			×			×	×					×		×	
Cisco: Connected Urban Development	×			×					×	×			×		
The Philips Center for Health and Well-being			×		×		×			×			×		

© Monaghan, 2012

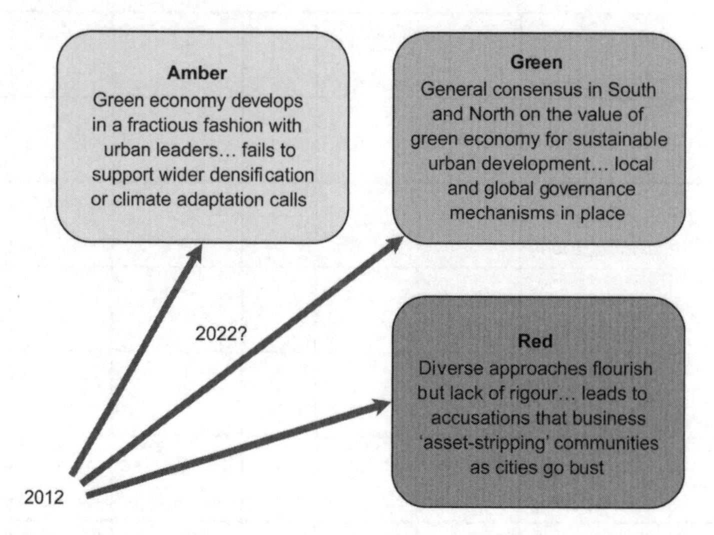

FIGURE 8.1 Future scenarios for green economy and urbanism? (© Monaghan, 2012)

problematic or at best, sub-optimal. However, the worst case scenario (red) is possible too, with weak governance arrangements between cities and the private sector leading to calls that companies are simply green washing their profit-over-principle approach, once again, as some were alleged to have done with eco-labelling in the 1980s and social auditing in the 1990s (and further to the paradigm shift of 'tomorrow's problems' referenced in Chapter 2). More optimistically, the best case (green) scenario is achievable too, if we learn from lessons of the past: that it is ok to be pro-responsible business but anti-weak governance. Here there is agreement amongst developed and developing nations that decarbonising the economy is possible in a way that raises the standards of living for the poor, with robust oversight arrangements in place at an early stage. A case in point is the forward thinking by city leaders in Quito (Monaghan, 2011g).

Ecosystem services in Quito (Ecuador)

Quito is the capital city of Ecuador and is located in the north of the country, in the Guayllabamba river basin on the eastern slopes of Pichincha, an active volcano in the Andes mountains. With a population of about 1.5 million, Quito is the second most populous city in Ecuador. Led by the Metropolitan District of Quito, the city is the second most important city to contribute to national GDP, and has the second highest per capita income. In 1978 Quito became one of the first UNESCO World Cultural Heritage Sites.

Located at an altitude approximately 2,850 metres above sea level, Quito's population depends largely on the water supply from the glaciers in the Andean mountains for irrigation and potable water. The glaciers of the tropical Andes are useful tools for studying variability in climate change and, worryingly, measurements since 1956 indicate a significant decrease in glacial coverage (UNEP, 2011a).

> 'There is a transferable lesson here for other places in terms of sustainability capacity building. SENPLADES, the national secretary for planning in Ecuador, established a countrywide enabling framework in the form of a National Plan of Good Living. This provided a much needed mechanism to build relationships with various actors, and for specific purposes,' says Dr Iván Narváez, FLASCO (Faculty of Social Science in Latin America) who led the technical team that during 2011 prepared a field report on the Environment and Climate Change Outlook of Quito (ECCO-Q) for UNEP.
>
> *(Monaghan, 2011g)*

Despite this, Quito offers a good example of the potential for developing markets that channel economic demand for water to upstream areas from where it is supplied (UNEP, 2009). The Fund for the Protection of Water (FONAG) was established in 2000 by the municipal government as a trust fund to which water users contribute. The proceeds from the fund are used to finance critical ecosystem services, including land acquisition for key hydrological activities. The users include farmers, hydropower companies, industries and households, who pay differentiated rates depending on use, with the water utility contributing the largest share (1 per cent of monthly water sales).

FONAG invests in both watershed management projects in river valleys and long-term environmental education, forestry training programmes and river basin management. These projects are carried out in partnership with community actors, educational institutions and NGOs.

Since 2009, FONAG has contributed to securing present and future water supplies for Quito, with more than 65,000ha of watersheds under improved management and with more than 1,800 people estimated to have received increased economic benefits through improved livelihoods and employment generation.

> The ECCO-Q makes a series of recommendations for future action on water amongst other climate change related aspects, given continuing pressures in relation to the poorest people lacking access to potable water or the use of contaminated water to irrigate crops, and cites the importance of FONAG's educational activities in helping to respond to this by promoting more sustainable water use practices. (This is one of a wider set of measures proposed along with a water pricing review and more of a multi-stakeholder approach to the stewardship of water given its use is a human right.) In concluding, 'Critical to success going forward is that national planning for wider climate change adaption and mitigation projects secures political agreement and is consistent with the development plan of the city' says Dr Narváez.
>
> *(Monaghan, 2011g)*

For further information visit http://www.unep.org/greeneconomy and http://www.fonag.org.ec (accessed 19 June 2011).

As is evident from the case of Quito its approach to the green economy is successful as it benefits the disadvantaged in a fair and reasonable way, is backed up by strong local governance through the local municipal authority and, ultimately, helps the city and surrounding areas adapt to climate change.

To realise this best case scenario on a mass scale would require key actors in local government and multilateral agencies to come together. Organisations such as ICLEI-Local Governments for Sustainability, the UN (UN-Habitat, UNEP and others) and OECD are beginning to grasp that they have critical roles to play in combining their expertise and influence to shape and monitor new 'CSR-urbanisation' and traditional PPP schemes alike (UN-Habitat, 2011) (A particular focus area over the next decade or so is the mid-sized cities of the East and West and notably China which will become the world's economic powerhouses by 2025: see Chapter 8.)

By 'stepping up' to this role, these agencies will also support responsible businesses who are doing the 'right thing' and whose positive contributions should not be hampered or go unrewarded. In addition, it will also help to define what the green economy means for a well adapted city in the South and the North alike. The end goal should be shared prosperity with urban resilience against future shocks, including the capability to learn, as this is universally desirable, avoids conflict and is ultimately more sustainable.

By intervening, the UN and its allies will no doubt face criticism. It will be labelled as being too controlling by those who see municipal authorities as too conservative when it comes to measuring the value-added the private sector can bring by taking over underperforming public services (Cohen, 1999). Others will consider that the UN's approach to the promotion of the private sector through PPPs and similar vehicles is somewhat ad hoc and is being done without the establishment of well-grounded criteria and procedures (Reed and Reed, 2006). And, whilst elements of both arguments have merit, the reality is that, in light of the banking and social care mismanagement (to name but two) by the private sector, with disastrous consequences in recent years, the same should not be allowed to happen with major urban infrastructure projects in our cities and our hope that the green economy brings a better future. Ultimately, appropriate 'checks and balances' need to be in place.

These checks and balances are something Zadek and Radovich (2006) refer to as 'governing collaborate governance'. Whilst citing mostly international developments projects, the transferable learning here is that the performance of such partnerships depends on how robustly they make decisions and their legitimacy to key stakeholders. This in turn depends on their governance structures and norms. A positive case in point for Zadek and Radovich is the Multi-Fibre Agreement (MFA) Forum (a global network partnership that operated between 2004 and 2010) to examine and address the impacts of the end of quotas for the garment and textile workforce in developing countries such as Bangladesh. Participants included retailers such as Gap Inc and Nike, international NGOs, multilateral organisations and trade unions. An executive committee made up of representatives from each of these sectors took operational decisions of behalf of the other participants, thus securing legitimacy, representation and participation. If this approach can work in textiles, why not other sectors too?

So, now having seen considerations of the interface between urbanisation and the green economy, it is necessary to turn to vested interests.

Conclusions

- The green economy is important, and is already 'open for business' from China to the USA.
- With no universal climate deal likely to be soon, the UN is turning its attention to the green economy, though there are questions over definition and governance, in particular in terms of the interface with the urbanisation agenda.
- Local leaders need to show courage in tackling societies' big challenges. This means shedding the right kind of caution when it comes to the green economy.
- Private sector involvement in the design and operation of major infrastructure should not be allowed to undermine the key principle of subsidiarity. This is about being pro-business but anti-weak governance.
- Decoupling carbon from growth must be done in such a fashion that the developing world and the poor benefit too. Otherwise, the transition to a green economy will fail. Governing collaborative governance is key to this. Multilateral agencies such as the UN and OCED need to be a lead facilitator here.

9

DECOUPLING VESTED INTERESTS

There is a misreading of economic resilience. It is not just about scale, but about scope as having more enterprises of the same can simply exacerbate a tendency to operate as a herd. What you need is a genuine diversity of organisations, including a range of ownership forms and industry sectors.

Ed Mayo, Secretary General, Cooperatives UK and Co-author, Consumers Kids (Monaghan, 2011h)

Ending unhealthy relationships

Think about these statistics for one moment. Between 1998 and 2008 the finance industry spent US$5 billion on lobbying. In 2010 approximately 200 of Barclays' senior staff earned US$844 million, while thousands of shareholders who had lent US $81 billion of equity capital earned just US$1,042 million in dividends (Bunting, 2011). By 2011, despite many banks benefiting from taxpayer bailouts in the UK, there was a US$3.5 billion shortfall in lending to small businesses promised as a condition of the loan; with renewed criticism that the banks that had caused the crash were also stunting recovery, in a single day alone protests at banks took place in 35 cities and towns (Toynbee, 2011). Box 9.1 contains further details of protests against banks and media companies.

Box 9.1 UK civil protests against banks spread to USA

UK Uncut – the activist-led NGO set up in late 2010 to campaign against public service cuts as a result of the bank bailouts and the failure of big business to pay tax – has successfully coordinated demonstrations from London to Hawaii (Hill, 2011). In early 2011, protests against banks involved parents and children transforming branches of the Bank of America and Lloyds into crèches, youth centres, classrooms, libraries, drama clubs, walk-in clinics and job centres. Protestors argued that these banks would have collapsed if it were not for the billions given to them by the taxpayer, and that it was their greed and poor management which caused the crisis, and that, as a result, ordinary people are paying the price through the subsequent cuts in public services (whilst bankers themselves are still awarding themselves big bonuses).

This kind of protest against irresponsible companies is not unique of course. Around the same time, the pressure group Avaaz rallied mass outrage on the Internet through an online petition against News International's takeover of BSkyB on the basis that this was bad for democracy and that the media group was *already* too powerful (Kingsley, 2011). Subsequently, in Summer 2011 following revelations that News International-owned newspaper *The News of the World* had hacked into voicemail messages (including those of murder victims and their families), Avaaz organisers targeted hundreds of the newspaper's advertisers seeking their support in a boycott of the title and within two months all but for four had cancelled their support. Founded in 2007, Avaaz is one of a new breed of so-called 'clickivist' digital campaign groups.

National government responses on regulatory reform have been varied. In the UK, for instance, the Independent Commission on Banking was established in 2010 to prevent a similar crisis to the global financial crisis of 2008/09 occurring in the future. The Commission has already recommended that banks be forced to ring-fence their high street banking businesses from more risky functions and to hold more capital, that is, at least 10 per cent; 7 per cent is currently required (Treanor and Elliott, 2011). This subsequently led the credit rating agency Moody's to put Lloyds TSB, Royal Bank of Scotland and Santander UK on a review for possible credit downgrades on the basis that banks may be left to collapse next time without tax-payer bailouts in the future (Hawkes and Wachman, 2011).

In contrast, the US Secretary of State has gone further and called for universally agreed global regulations – to avoid a race to the bottom (i.e. that banks are able to move to other markets where regulation may not exist). If these regulations were ever in place, there are many ideas about what or how they could work; for instance many groups in the UK (including Oxfam and War on Want) have called for a 'Robin Hood Tax', whereby a levy is placed on financial transactions (Treanor, 2011). Another potential way of 'recycling' these funds could be to support the most needy, such as pursuit of the MDGs.

However, even if such a tax or levy were applied surely it would just be 'tinkering around the edges' of a failed banking system. Why would we stay wedded to such a system that brought the world to its knees only 2 years ago? If unfettered free trade does not make countries better off, why do we appear to be a slave to the markets? (Chang, 2010). In short, is all hope lost when it comes to reining in the negative forces of the big banks? No, possibly not.

Maybe a new economic model could yet remedy this system. Korten (2011) set out a vision of what a 'new economy' could look like, starting with how differently governments could have dealt with failing banks during the global financial crisis of 2008/09. Instead of bailing them out, they would have taken them over, negotiated financial settlements with creditors at dramatically reduced prices and restructured the big banks to hive off parts as independent locally owned community banks. At the same time, any financial institutions devoted to activities considered unproductive to the cause would have been closed. The bailout money therefore would have been directed to keeping ordinary people working and in their homes.

Orthodox economists, not surprisingly, counter this by stating that the sub-prime mortgage crisis in the USA (which subsequently reverberated around the world) arose from earlier and similar liberal interventions as proposed by Korten – notably, the banks being forced to lend to the poor and profligate governments rather than unchecked global finance. As such, they maintain that the solution to the crisis created by an excess of debt cannot be remedied by creating more debt through a Keynesian spending stimulus.

Other liberal economists, however, such as Joseph Stiglitz (Elliott, 2011a) counter that excessive austerity measures lead to even higher unemployment, placing pressures on wages and therein overall demand in the economy that, in turn, dampens any sustained recovery.

The opportunity to mobilise the new way of thinking described by Korten still remains valid, with an ongoing lack of regulation to curb speculators such as Barclays who it is alleged were pushing up food prices in 2011 which was putting staples out of reach for the poor (Inman, 2011). A more detailed narrative of how to return to 'good banking' is detailed in Box 9.2.

Box 9.2 Good banking gone rogue?

Calls for drastic reform of the financial markets have been made on both sides of the Atlantic on the basis that banks have completely lost their way and no longer serve the public good. Whilst the 2008/09 global banking crisis brought this into particular focus, the ability of rating agency Standard and Poor to downgrade the USA's national credit worthiness in 2011 provided another timely reminder too.

In the USA the New Economy Working Group (2011) argued that since the 1970s ordinary people have yielded the power to control the creation and allocation of money to a small group of bankers in Wall Street who are only accountable to themselves. And, further, that this institutional failure led in

2008 to Wall Street plunging the US economy into the worst crisis since the Great Depression, and, whilst the banks received a generous public bailout and quickly recovered to reward themselves with record profits and bonuses, ordinary people are still hurting through loss of employment and homes. Indeed, some of the banks that held foreclosed homes were the same lenders that had received public bailouts from taxpayers whose livelihoods were destroyed as a result.

The New Economy Working Group (2011) concludes that to 'liberate America from Wall Street rule' we need:

- to reverse the process of banking consolidation and build a national system of locally based and accountable financial institutions devoted to building community prosperity by breaking up the big banks and introducing tax incentives that favour non-profit cooperatives;
- to create a state partnership bank in each of the 50 states to serve as a depository for state financial assets, which keeps locally held money through partnerships with local community banks providing loans to local homebuyers and businesses;
- to restructure the Federal Reserve to limit its responsibility to managing the money supply, subject it to federal oversight and public accountability, and establish specialised regulatory agencies;
- to create a Federal Recovery and Reconstruction Bank to finance critical green infrastructure investment, with projects designated by Congress;
- to renegotiate international trade and investment rules to secure greater national ownership and self-determination, for instance holding corporations operating in multiple countries accountable for compliance with domestic law in each country of operation;
- to implement regulatory and fiscal measures that secure the integrity of financial markets, such as rendering speculation illegal and unprofitable.

Proposals for these revisions in the USA came at the same time and have common features with those being developed at a similar moment in time through the 'good banking summit' convened in the UK by the New Economics Foundation and Compass (NEF and Compass, 2011). The summit was organised due to fears over an apparent return to business as usual in the financial sector despite an independent review of its practices and the growing public dissent about the poor conduct of banks in the wake of the 2008/09 banking crisis and the subsequent UK economic recession. The summit highlighted examples, such as the loans of the top 10 UK banks are 450 per cent of national annual output whereas back in 1960 it was just 60 per cent, to demonstrate why the system is fundamentally flawed.

The NEF and Compass (ibid.) concluded that a number of key changes are needed for a return to 'good banking':

- break up the banks, in order to ensure localisation and competition for community and business needs;
- rein in pay, risk and perverse incentives in banking;
- make money creation and supply more democratic to meet everyone's needs;
- address the influence of the banking lobby;
- establish a Post Bank and Green Investment Bank to ensure fair access for all and help build a low carbon economy;
- assess the assessors, correcting the role of accounting and credit ratings agencies in the banking failure.

In summary, they found that a key focus of any review should be of the associated systemic risks, given the current crisis has resulted in 800,000 job losses which has left the UK 10 per cent poorer (ibid.).

Realistically, to be most effective efforts to decouple vested interests must be translated at the sector level as equally as at the meta-market level, whether it be in the home mortgage industry or environmental services. Take for instance attempts to secure market traction for recyclers in Latin America.

Bringing informal waste collectors into the mainstream in Asuncion (Paraguay)

In parts of Latin America (such as Argentina, Bolivia, Brazil, Colombia, Paraguay and Peru), 4 million people are 'informal collectors' of recycling and are responsible for recovering up to 90 per cent of all recycled consumer products. Yet these collectors only receive a minor percentage of the value they generate due to a lack of regulatory and market mechanisms. In addition, they operate in working conditions which threaten their health through handling dangerous materials (Green Economy, 2011).

In order to help solve this problem, the Inter-American Development Bank (IDB) developed a novel partnership with municipal authorities and Coca-Cola. The pan-Latin America project brings together recyclers, consumer product companies, local authorities and NGOs to discuss and agree plans to incorporate informal workers into local value chains.

As well as being a smart intervention to build capacity to help many impoverished people living at the margins of society, this is also a powerful example of the green economy working well in the South. Just as interestingly, the initiative to bring informal waste collectors into the mainstream works better without it being referred to as a 'green economy' project. Given that, at its heart, this is an initiative enriching the Southern poor by shifting power *to* them and *away* from Northern big business operating in their countries.

Strength through diversity as well as devolution

At the height of the banking meltdown in 2009 the Bank of England's Executive Director of Financial Stability, Andrew Haldane, published a seminal paper in which he used global fish stocks, the spread of diseases such as SARS and the deforestation in the Amazon to explain why the system failed so spectacularly. The key analogy was that some failures, whatever the concept, are due to a common vulnerability: an absence of diversity (Haldane, 2009).

Haldane's argument is that in the ten years leading up to the banking crisis, the financial sector became larger and more complex whilst also increasingly homogeneous. Mutual organisations became banks, commercial banks were allowed to trade in investment banking and investment banks were allowed to establish hedge funds. As a result, the financial system, like any ecosystem, including those for plants, animals and oceans, became less disease resistant as it became less diverse. When the environmental factors took a turn for the worse, the homogeneous characteristics of the financial ecosystem meant it was exposed to failure – it panicked and seized up. The tipping point was then reached the moment when only fractionally more irresponsible trading occurred but it caused irreparable damage because the financial system was no longer resilient due to its homogeneous nature. For any new financial ecosystems, whether mutuals, commercial banks, investment banks, state-owned banks or banks dedicated to social or green investment, to be resilient they need a richness of diversity (Elliott, 2011b).

A call for greater diversity in banking provision is echoed in electricity provision too, namely long-standing monopolistic energy distribution networks are inhibiting the move to decentralised renewable energy generation as detailed in Box 9.3.

Box 9.3 Time to break up the monopolistic energy distribution networks?

The Green Alliance (2010) – a leading environmental think tank – argues that technological and operational changes in electricity transmission and distribution networks are critical to meeting carbon reduction targets and combating climate change. Despite privatisation of the electricity market in the 1990s, competition only exists amongst generators, and networks remain regulated monopolies. They argue that we need to create competition in the networks, particularly at the distribution level, to lower costs and stimulate innovation and offer new low carbon services.

For example, in the UK, connecting to the network has become a significant barrier to progress of renewable energy projects, say Green Alliance, with a queue of 60GW worth of new generation capacity seeking connection to the grid. Connection dates now being offered are as late as 2023.

Proposed options to introduce competition include making the existing network distributor allow new entrants to use existing networks or the development of new networks in parallel to the existing ones. In a riposte to those who assume that developing more than one network in any single area is uneconomical, Green Alliance cite the experience in the telecoms industry which shows that, even if a very small market share is gained by new entrants, incumbent operators make enormous attempts to innovate and provide new services to ensure they do not lose custom.

Just as important is the advancement in the uptake of distributed heat and power schemes to help support the roll out of so-called 'smart grid' systems. As well as enabling local generators, including households, to sell surplus energy back to the national grid, this allows residents and businesses to recharge plug-in hybrid electric vehicles and helps building owners or managers make use of smart meter display units to better control energy supply and demand. One successful local authority example is Boulder, USA (Monaghan, 2010).

In response to this challenge, the UK energy regulator Ofgem launched an ambitious US$815 million low carbon networks fund in 2010 to 'kick-start' the radical change that the electricity networks need to make the transition to a low carbon economy (Ofgem, 2010). However, disappointingly, the first tier of funding was only made available to the incumbent distribution network operators.

Drawing strength through diversity as well as adversity is echoed at the municipal level too. Take for instance the following example of community land development in Madison, USA, where a community land trust (CLT) – a non-profit company or charity which acquires parcels of land in a geographical area to ensure the permanent affordability of any housing or other development – was set up to great effect (Monaghan, 2011i).

Madison community housing (USA)

Troy Gardens in Madison, Wisconsin, is a mixed-income homeownership project, organised as a condominium, using a co-housing model of community development. Developed by the Madison Area CLT, it includes 30 units, one-third of which were priced below the market for low-to-moderate-income families (Home Depot Foundation, 2008).

With Troy Gardens residents as key decision-makers, the homes are set in 31 acres, with over two-thirds permanently protected green space for community gardens, an organic farm, restored prairie and nature trails. All homes were designed and built for durability and energy efficiency to minimise operating costs (e.g. windows, lighting, appliances, air conditioners), meaning they received Wisconsin's Energy Start certification. Additionally, the buildings are oriented and designed for photovoltaic (PV) production of electricity and have thermal panels installed on the roof.

To foster a sense of community, all homes are clustered around courtyards, with front doors and porches facing other homes. The homes are also situated so there is easy access to transportation and within walking distance of schools and shops, to minimise car use and to promote healthier lifestyles.

'The great thing about co-ownership is that there are elements of "I" and "we" about it. You can have a stake in the house but the land is shared' says Pat Conaty, Fellow at the New Economics Foundation.

Despite these successes, trusts like this face big challenges in getting projects started. That is, the inability to get banks to lend because they do not or will not understand a trust's revolving fund model, whether competition for land is fierce (especially in inner city areas where it is prohibitively expensive or deemed more desirable by a developer for the buildings to sit idle until a 'better offer' comes along) or where the local authority's planning framework is not enabling – this can be particularly tricky if the CLT proposal is for a complex mixed-use development.

Yet, paradoxically, the advantage of limited or shared equity cooperative housing is essentially practical as these forms of housing benefit from several cost savings (Conaty et al., 2003), including:

- reduced interest on borrowing costs for new housing acquisition as the cheapest limited equity form of borrowing is corporate, not individual;
- best value in professional management fees as tenant participation in management can save on some overheads and offer continuous pressure for efficiency;
- simplification of tenure; a repairing lease and equity shares can be assigned through a transparent formula.

Such initiatives also have the added advantage in supporting people to get on the property ladder by gradually building up ownership of property. In doing so it stimulates social mobility, thus lowering the cost burden of welfare support for municipal authorities. Conaty concludes:

> Homes for every deserving person is not a new concept. It has been done before, many times, lest we forget 'Homes for Heroes' in the UK and the USA after the war. Why not again for key workers like nurses or teachers just starting out in life but who are currently excluded from entering into the property market?
>
> *(Conaty et al., 2003)*

For further information visit www.neweconomics.org/ and http://www.troygardens.net

Whilst critics may argue that CLTs have failed to prove that they are scalable, the riposte is that this is rather the system stacked against them doing so. Banks should be forced to lend, developers who sit on idle land should be thrown off and local authorities' urban development plans should include the right to build.

Neither Madison nor housing are unique areas or sectors in regard to drawing communal strength through diversity. Take for instance what Reed and Reed (2006) term new 'public–public partnerships', such as this water partnership in Santa Cruz.

Public–public water partnerships in Santa Cruz (Bolivia) and Finland

In the Bolivian city of Santa Cruz water services are managed by Saguapac, the world's biggest cooperative. It provides water for 80 per cent of households across a population of 1.25 million people and is understood to be more efficient and transparent than its private sector peers (Reed and Reed, 2006). Every user is a co-owner and has voting rights including electing the company board, as well as evaluating the quality of service to determine the priorities for improvement. A social tariff is in operation to match people's consumption needs with their ability to pay.

There have been similar successful experiences in rural Finland (UN DESA, 2005), where over a thousand water cooperatives serve farm businesses and villages. Whilst licensed by the government and allotted a limit to the amount of water they can extract, the cooperatives have complete control over pricing. This means they can offer favourable rates to their members because pricing decisions are not influenced by fluctuations of the market. Finnish water cooperatives also have the network benefits of partnering with other regional associations: if, for instance, the water quality in one area is not sufficient, the cooperative can buy water from a neighbouring cooperative-owned water network. This ensures continued low prices as well as supply dependency.

So, though a 'public–public' partnership is one potential formula for success, it is important to strike the right balance. Social commentators tend to either declare all forms of communal management inefficient and in need of privatisation or conclude that local management has superior user intelligence than big business (Vaccaro et al., 2009). However, upon closer examination smarter but less intervention by private, public and community actors may be the wisest course of action. Perhaps easier said than done; this is something which the next chapter considers.

Conclusions

- A failure to end unhealthy relationships will expose a fragile society to collapse.
- Strength can be garnered through diversity. Diversity relates to variety in the forms of the organisation within a sector, as well as the types of sector and the robustness of these networks or relationships.
- Citizen-led provision offers the double benefit of devolved power and a strengthened resistance to ecosystem failure.

Part Four
Transition from unstable to resilient societies: hard to make, hard to break

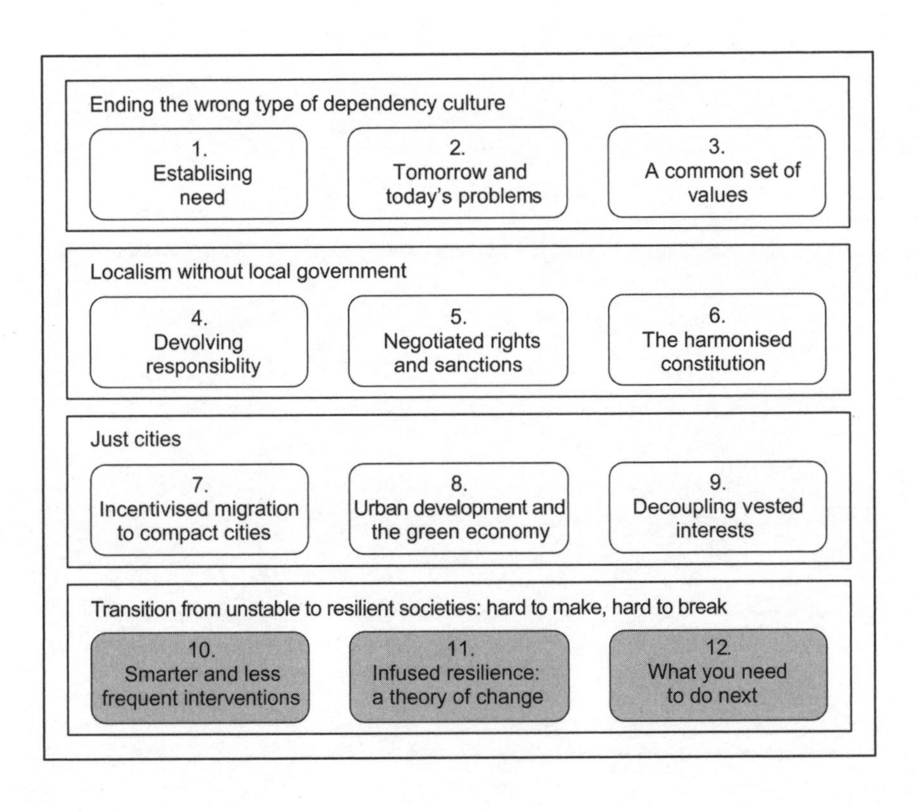

10

SMARTER AND LESS FREQUENT INTERVENTIONS

We need to overcome 'discipline apartheid'.

Professor Mark Swilling, Sustainability Institute, University of Stellenbosch, South Africa (commenting on the required competencies for city leaders to deliver sustainable urban development) (Monaghan, 2011i)

Resource flows

The fourth and final part of this book is intended to show how leaders can manage the sustainable transformation to a more resilient way of living and working, as the world recovers from an age of austerity that is sustainable, practical and resource-effective.

You will recall that, so far, this book has called for an end to the wrong type of dependency culture, explored the notion of localism without local government and focused in on just cities. Fundamental to this is cautioning against single-issue policies, well intentioned or not, which go against the grain of the systemic change required.

Given all of this, a key consideration remains the issue of materials use ranging from utilities to construction. The cold hard truth, as we know, is that the mismanagement and over consumption of natural resources is unsustainable And, so, to deal with this we firstly need to understand the complex flow of materials in and out of our societies; and secondly we also need to determine how best to govern them.

Whilst the long-term historical trajectory of resource prices has been a downward one during the 20th century (with only odd periods of price peaks), this is not a proxy for the future (International Resource Panel, 2011). Markets are often guilty of discounting the high social and environmental cost of scarcity, which is especially risky if it leads to corrective action coming too late in the case of life essential commodities like food, UNEP's International Resource Panel calls attention to the decoupling of human well-being from resource consumption by linking local or national development strategies to resource flow strategies. The IRP analysis shows that by 2050 the level of resources used by each person each year will need to fall dramatically – to 5–6 tonnes – in order for us to live within our environmental limits. Consumption levels vary wildly, with some developing countries still below this level (e.g. India at 4 tonnes per capita) whilst some developed economies are as high as 25 tonnes per capita (e.g. Canada).

Even in two countries that are making explicit efforts at decoupling human well-being from resource consumption – Germany (i.e. National Strategy for Sustainable Development) and Japan (i.e. Sustainable Society Policy) – where it would appear domestic resource consumption shows stabilisation or even a slight decline, further analysis shows that many goods contain parts that have been produced abroad using major amounts of energy, water and minerals. So some countries are managing the problem of high resource intensity by exporting the problem elsewhere!

Additional studies (Environmentalist, 2011b) have also found that emissions from imported goods exceeded by five times the emission savings made by industrial nations between 1990 and 2008. This means that these countries' national carbon reduction targets are increasingly less credible as they fail to include emissions arising from overseas sourced goods.

Despite this huge conundrum, IRP points out that rapid urbanisation combined with technological and systematic innovation offer an historic opportunity to reduce our 'metabolic rate'. That is, the processing of resources passing through the city can been seen to constitute a city's 'metabolism,' as described in Box 10.1.

Box 10.1 The city as a super organism

Swilling et al. (2011) state that to understand the interaction between cities and the environment it is useful to consider the flows of natural resources needed to sustain them. Flows of water, energy and other commodities enter a city and, while some remain to constitute the fabric of the city through construction materials, most are changed through human use and exit as waste to landfills, watercourses and into the atmosphere. A city's reliance on inputs for survival can be compared to a super organism, where roads, railways and watercourses act as veins, food markets operate as a stomach and waste landfill sites are like the digestion system.

Swilling also argues that for this super organism to operate as effectively as possible and be able to deal with complex changes, city officials must overcome 'discipline apartheid' (which is about different disciplines such as accountancy, engineering, regeneration or environmental services coming together to co-develop solutions).

An interesting example of a city attempting to reduce resource consumption is Lima.

Wastewater reuse for irrigation in Lima (Peru)

Rainfall in Lima is very low, with the city almost entirely dependent on its surface and groundwater supplies. Worse still, climate change is likely to lead to an increase in temperatures causing the glaciers to melt and resulting in increased water scarcity (ICLEI and UNESCO-IHE, 2011).

In response, multifunctional use of water sinks (a natural phenomenon or human-made facility where water gathers or is collected) has been utilised by local government through their involvement in the SWITCH project (led by ICLEI – Local Governments for Sustainability and UNESCO-IHE) which includes recycling treated wastewater as an alternative for irrigation water, thus saving clean water being diverted from drinking use.

Returning to the IRP's commentary on Germany and Japan it is perhaps ironic that the well-intentioned 'Eco^2 Cities' programme by the World Bank (2010b) has an exclusive focus to help cities in developing countries achieve greater ecological and economic sustainability as cities in *developed* countries need to do their bit too, regardless of whether they are shrinking or not (given the city migration patterns described in Chapter 8 which will involve depopulation in the North as a result of the shift to the South and East).

So, whilst applauding IRP's analysis of resource flows, one limitation is the lack of examination of the dependencies of resource flows *between* cities. This is explored further in Box 10.2.

Box 10.2 'Insuring' against cities dependencies

Leading global insurance company Swiss Re's work on the economics of climate adaptation has led them to develop a framework to help local councils make the most informed decision regarding when to invest in preventative measures against climate change related emergencies such as flooding or drought, and when it is more economical for them to take out 'damage' insurance.

Swiss Re's research in Hull (UK), Miami (USA), Maharashtra (India) and Georgetown (Guyana), amongst other cities, presents a strong case for immediate action – that is, it is cheaper to adapt now than wait for the damage to occur (Bresch and Spiegel, 2011).

This linear risk assessment is important and helpful, but what about the risk posed by trade between cities in terms of resource flows? Of particular concern are the 100 or so mid-sized cities in the East and South (like Guiyang in China or India's Surat) which are expected to become the new economic powerhouses over the next 15 years. As we saw in Chapter 7 these are cities producing 1/3 of global GDP, but are also within nations which may face major problems operating within their environmental limits (e.g. with water risk sensitivity) now or in the very near future (Maplecroft, 2011).

Indo-China trade in 2010 was US$60 billion and is targeted to rise to US $100 billion by 2015 (BBC, 2010). So, if the trade between Guiyang and Surat (water scarce areas) is in water-intensive industries such as agriculture or textiles, then, unless one or both of them has a diverse supplier or export base, their resilience could be fragile.

More than this, given that these cities will be a key target for major inward investment, could their failure precipitate another world recession? Drawing upon separate but transferable learning from Castells (1990), which concludes that power now rests in networks and city populations lose control if they fail to understand the clash between flows and places, it can be argued that city planners and insurance firms alike need to grasp that current understanding of risk and resilience is dangerously narrow if it does not include resource flows between cities.

Harnessing the positive power of markets and people

To make smart interventions on resource flows necessitates the better harnessing of both the markets and people. This will require the redirection of financial investments away from speculation and into more productive capacity for sustainable urban development (UN-Habitat, 2011), be it new major low-carbon infrastructure projects (e.g. rapid surface transport) or lending to small businesses and more diverse forms of organisations such as cooperative enterprises (which stimulate innovation and empower and inoculate us against the disease of financial collapse, as we have seen in Chapter 9).

Smarter finance also means keeping as much of the money generated from transactions in priority or poor areas as local as possible (so these resources are reused as money is respent locally), as opposed to it being repatriated to outside of the said area or to wealthy actors (NEF, 2008). An example of this practice comes from Manchester, a large city in the UK.

Local buying to save money and improve communities in Manchester (UK)

An independent study by CLES (2011) of Manchester City Council's annual US$572 million procurement spend determines how much is spent locally. CLES concluded that 51 per cent is spent on organisations within the city, with 25 cents in every US$1 received by top suppliers respent locally (CLES, 2011), which compares well to local authorities in Swindon (16 cents in every US$1) and West Lothian (18 cents in every US$1).

The learning here is that, while EU procurement rules can sometimes prevent local procurement, suppliers could still be asked by councils to tender their bids to include efforts to nurture local economic development, for instance by offering apprenticeships.

In an age of financial *and* natural resource austerity, how do urban leaders begin to select the best intervention from the choices available? To assist decision-makers build capabilities in this regard this section shares a number of tools in three easy steps.

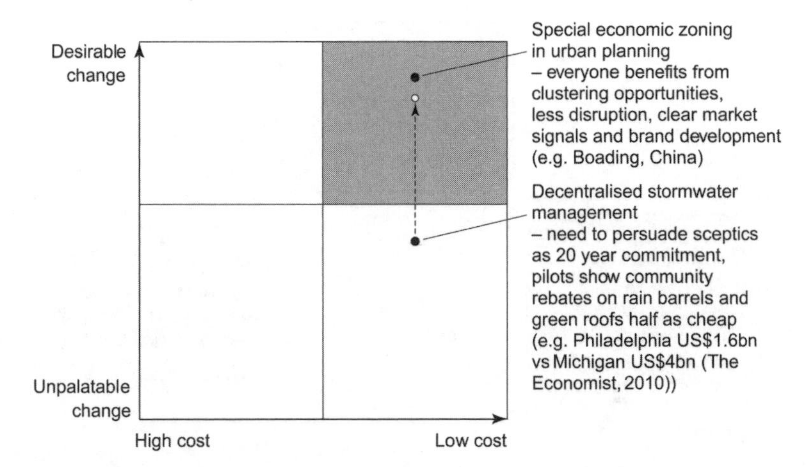

FIGURE 10.1 Intervention matrix (for smarter and less frequent actions) (© Monaghan, 2012. Adapted from Monaghan, 2010)

Firstly, as Figure 10.1 shows, an intervention matrix is helpful in whittling down a long list of potential solutions on material issues (with 'material' referring to importance not resource). Here, the materiality rating may be filtered by criteria including major natural resource flows, strategic alignment, elected member and leadership priorities, available finance, value for money, compliance with environmental or social legislation, better management of risk, brand protection, delivering expected services or joint ventures with like-minded partners.

Once materiality is determined, the matrix plots desirable change against acceptable cost, with the most promising interventions in the top right-hand quadrant. Moreover, in the absence of effective regulation, a compelling business case will often need to be made to take voluntary action, which is where this tool comes in. Desirability here may be impacted by an intervention being on close-to-home issues or values, perceived fairness, existing habits or sheer practicability. Ability to bear costs is again context specific, but will include invest to save options (i.e. cost prevention). This point is illustrated using the examples of special trade zoning versus bottom-up solutions to deal with stormwater in China and the USA respectively.

Secondly, to realise these benefits, however, one must be open to new ways of doing business. That is, embracing the concept of 'green concentrate', which involves engaging with partners in a way that optimises financial and natural resource use, but is ultimately about removing choice too (as illustrated in Figure 10.2). An excellent case in point here, as referenced earlier in Chapter 5, is the 'carrot and stick' citizenship schemes which 'edit' resident choices on waste minimisation in Neustaudt an der Weinstrasse, Germany.

Thirdly, just as crucially in formulating the business case, in order to make use of limited resources, municipal authorities need to use all existing pathways at their disposal. Whether local authorities control or influence an event (e.g. planning policy

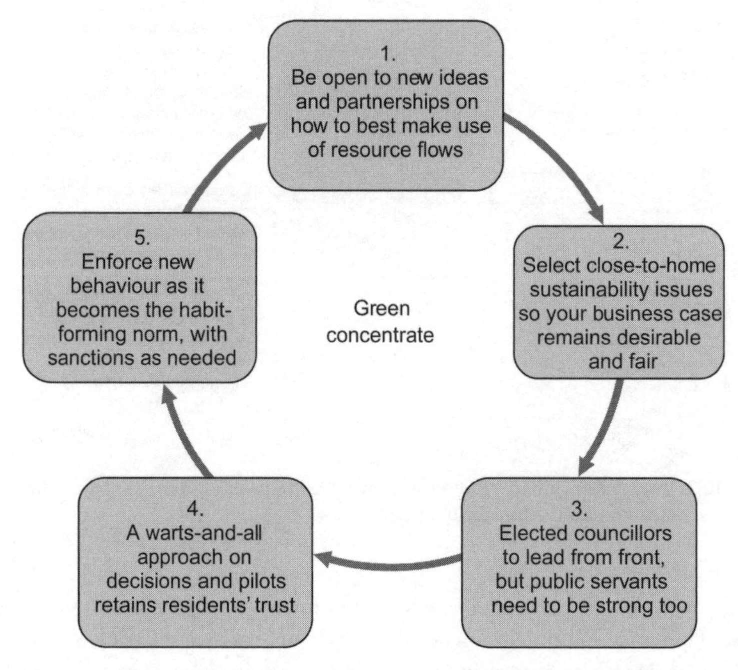

FIGURE 10.2 A framework for excellence (by local government in an age of financial and natural resource austerity) (© Monaghan, 2012. Adapted from Monaghan, 2010)

and consumer choices respectively), they need to mobilise all available levers to achieve the desired impact.

So, for instance, if a primary aim is to create local employment opportunities, this will require the economic development department overcoming 'discipline apartheid' and working with colleagues in corporate assets and resources to ensure that the public pension investment policy supports investment in local projects. Or that the corporate procurement policy ensures that as much of the council's spending contracts go to domiciled employers and workers (whilst observing fair play rules). This should be in addition to liaising with colleagues in environmental services to ensure that any resulting economic transaction is decoupled from rising carbon emissions (see Figure 10.3).

In making the best use of a multitude of key pathways, public servants will need to be brave and challenge colleagues who prefer to work in silos. The normalisation of such behaviours within an organisation becomes critical here (see Figure 10.4).

This is, however, the beginning of making a smart intervention given that, as we now know, a systems-thinking approach is fundamental to success and to avoid unintended consequences. To bring this point home, let us take one example of what some observers consider to be 'getting it wrong'. Geo-engineering is being investigated by the UN Intergovernmental Panel on Climate Change (IPCC) as one solution to climate change – such as sending huge mirrors and sulphate aerosols into the stratosphere to reflect sunlight back into space (Vidal, 2011b). Opponents argue that this would undermine energy security as it prevents sunlight reaching solar panels,

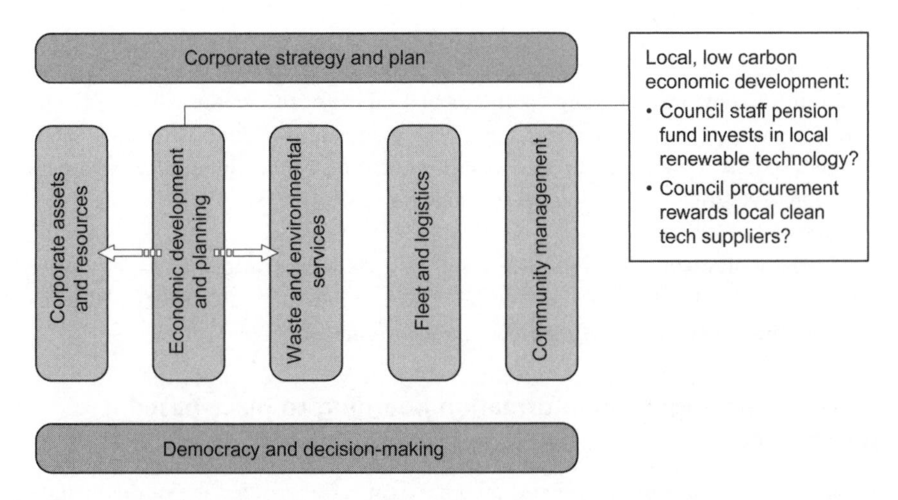

FIGURE 10.3 Key pathways to smarter outcomes (© Monaghan, 2012. Adapted from Monaghan, 2010)

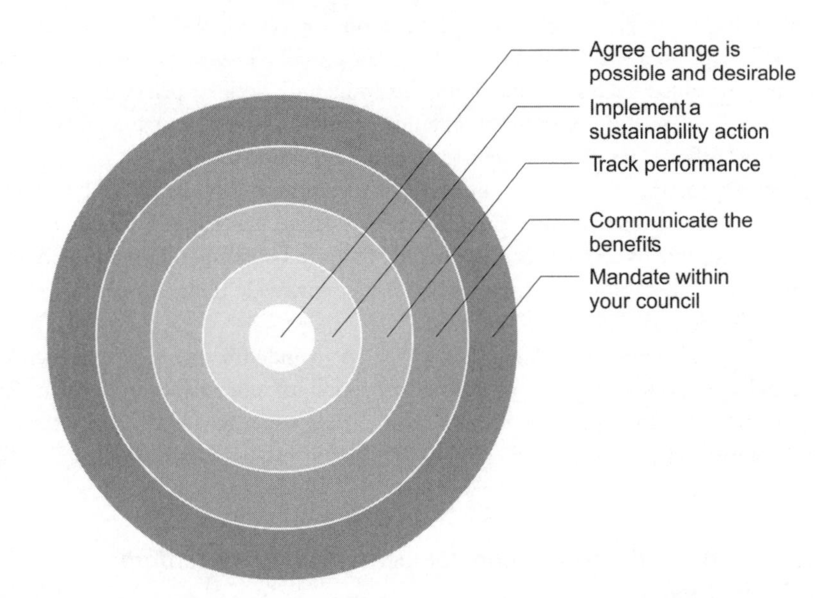

FIGURE 10.4 Normalising action on sustainability (© Monaghan, 2012)

which in turn prevents us weaning ourselves off the fossil fuel addiction (the same addiction which caused the problem in the first place) (Kintisch, 2010).

In contrast here is an example of possibly getting it right, with a multi-faceted approach to managing resource flows in Changwon.

Declaration of an environmental capital at Changwon (Republic of Korea)

In 2006 the local government in Changwon set out an ambitious 11-point pledge to transform the city into a leading environmental capital of the world.

Key commitments in Changwon's five-year master plan include a resources management system as well as cleaner air, water restoration, a green transportation system and more sustainable energy sources (Changwon, 2006).

Whilst this is all useful, how can these complex systems be managed in a robust way? That is, how can local government translate systems thinking into strategy and technical operations whilst also being accountable to ordinary people?

Systems thinking: from information hoarding to place-based governance

So what exactly is meant by 'systems thinking'? An early recorded use of the method was after the Second World War (Middleton and Seddon, 2010) in the context of the survival of ecological systems, and as with resilience, the term has been subject to a variety of interpretations and uses.

Building on earlier work by Forrester (1971), Meadows (1997) argues there are places within a complex system – a corporation, an economy, a living body, an ecosystem – where a small shift in one thing can produce big changes in everything else. These so called 'leverage points' allow us to identify the best places to intervene in a system and include: resource constraints, buffers and other stabilising stocks, information flows, rules of the system and power distribution.

Specifically in the context of public service managers, Middleton and Seddon (2010) go as far as to say that, in the UK and elsewhere, the failure of thousands of practitioners to understand this concept has led them to be engaged in the imposition of 'command and control' target setting that has simply made things worse, as well as costing millions of dollars.

Things are beginning to change, however. A wonderful example of systems-thinking by a municipal authority is the instigation of sustainable food chains in Amsterdam which addresses a number of cross-cutting issues related to food security, climate mitigation and adaptation, obesity, improving rural–urban relations and city competitiveness in one (Monaghan, 2011j).

Food, urban–rural linkages and competiveness in Amsterdam (the Netherlands)

Amsterdam is the nation's capital and largest city with a metropolitan population of over 2 million. The city is in the province of North Holland in the west of the country, and is the financial and cultural heart of the Netherlands with many large Dutch institutions headquartered there including Philips and ING. The city is also characterised by rural functions such as farming which contrast sharply with the urban areas around it (ICLEI, 2008).

This is the world's second largest exporter of agricultural produce (40 per cent of Amsterdam's ecological footprint is made up by the provision of food, comprising dairy, meat and other agricultural produce, but excluding transport). Yet, local produce is still a niche market and in addition unhealthy eating habits and lifestyles are causing considerable negative impacts on the city's residents; for instance 70 per cent of health problems are related to food, with 45 per cent of the people of Amsterdam overweight and 14 per cent of these classed as obese. At the same time, Amsterdam's problems are set to increase as the Food and Agriculture Organization (FAO) predicts agricultural production will need to increase by at least 70 per cent to meet demands by 2050, based on population growth and food consumption, with most estimates indicating that climate change is likely to reduce agricultural productivity and production stability (FAO, 2010).

Since 2006/07 the municipal authority of Amsterdam has placed strong emphasis on the importance of food for sustainable urban–rural development.

'The strategy – as defined in 2006 – was to combine, connect and scale-up initiatives in the realm of sustainable regional food chains; rethink urban–rural relations and promote healthy food, diets and lifestyles in the region. Basically this bottom-up strategy is still being applied in our municipal departments,' says Pim Vermeulen, Senior Planner in the Amsterdam Physical Planning Department. 'Gradually we better understand how all the parts in the food cycle are interrelated and how we could shape coherent policies – from primary production to processing, distribution, wholesale marketing, preparation and storage, consumption and waste.'

Amsterdam plans to achieve this through a holistic mix of more intelligent interventions, including:

- planning protection for 50 per cent of land which is rural and used to grow food, e.g. potato, dairy (this also helpfully acts as a flood defence);
- using the council's spending power by procuring locally produced food in council canteens to stimulate demand amongst local shops to stock items from local farmers;
- nurturing neighbourhood level urban food growing projects such as community vegetable gardens and temporary cornfields which make use of local knowledge, as well as school vegetable plots and allotments;
- stimulating innovative food-growing techniques in the urban environment (e.g. 'Plantlab', a method of indoor growing with LED-lights);
- building capacity through NGOs, e.g. those working with schools and further education colleges to deliver courses on how to cook meals that meet nutritional needs and are organic.

A multi-stakeholder food strategy operated until 2010. Budget cuts as result of the recession meant the council could not continue to support the programme in the same way in future years. Despite this many former participants continue to co-operate on specific local projects. Vermeulen again:

> We particularly welcome alliances and bottom-up initiatives that can be scaled up. Not only does the council not have the resources to always lead, other

partners may be better placed to do so anyway, especially local enterprises, community groups or NGOs.

Citing obstacles to be overcome, Vermeulen highlights that:

> The challenge will be to maintain this momentum, but also get more clarity on the role of the national government. How can we attain a better coordination of national, regional and local food policies, in terms of food education and the organization of food cycles that exploit the potential of local natural resources while minimally affecting the environment?

There are large rewards at the end of this – for the enhanced competiveness of Amsterdam, as well as ensuring Amsterdam's food resiliency – by encouraging diversification and innovation of urban–rural relationships, thus promoting future economic development in peri-urban areas and helping the metropolitan area meet the dilemmas of shifting changing global food markets and demographic developments.

For further information visit www.amsterdam.nl

The discipline of systems thinking is beginning to mature to the extent that many municipal authorities have created the new role of 'Chief Information Officer' (CIO), who is tasked with deriving better business value through smarter use of information. The CIO's duties are more than simply those of an Information Technology (IT) Director, as they involve checking the system works, devising plans aligned to key leverage points and then taking action to maintain or improve this complex system.

The rise of the CIO is not a surprising development of course given, for instance, that infrastructure planning is key to economic development strategy for city leaders and the use of information-based instruments is increasingly helpful to facilitate alignment of these plans to wider environmental and social concerns. Examples of information-based instruments include environmental management systems, natural resource or carbon budgets, biodiversity indices and geographic information systems (UN-Habitat, 2011). These information-based systems are also being used to encourage engagement and grow awareness. There are numerous examples of this in action from Canada to Colombia.

Creative use of technologies in the Jätkäsarri quarter of Helsinki (Finland), Ontario (Canada), Medellin (Colombia) and Paredes (Portugal)

According to engineering firm Arup (2010), cities are real-time systems but are rarely run as such. Whilst, in the past, several cities have used information and communication technology (ICT) to improve municipal performance – ranging from mobility to e-government services – a new generation of leading cities have pushed this boundary even further. These so called 'smart cities' are seeking to realise the added value by moving from departmental solutions to an area-wide approach:

- Jätkäsarri, Helsinki: an urban development which features a wide range of pervasive informatics strategies and services aimed at a dramatic reduction in the community's

carbon footprint. Services will include in-street displays that report on personal and civic footprints in real time, next-generation smart meters for apartments, and displays and services that highlight patterns of production ranging from urban agriculture to knowledge-based work (Arup, 2010).
- Ontario, Canada: a blackout in 2003 – the largest in North American history – arose through a combination of software failures, overburdened transmission lines and outdated control equipment. The province of Ontario responded with plans to upgrade its electricity supply infrastructure by 2025, so that the new grid could detect problems and automatically contain them. Subsequent proof-of-concept pilots on smart meters between 2004 and 2006 showed that 75 per cent of the population could benefit from a 6.5 per cent energy reduction. This culminated in the Energy and Green Economy Act 2009 that promoted the roll out of smart technologies to increase energy conservation, better integrate renewable energy sources, as well as the creation of 50,000 jobs (Rustin, 2010; Sarchet, 2011).
- Medellin, Colombia: in 2004 the city's cable car system began carrying residents from its suburbs to the centre, feeding directly into its metro system, reducing a typical journey from 2 hours to just 7 minutes (Morris, 2010). Integrating the city's poor was a key goal of the smart project in an attempt to reduce gang violence. So crèches and libraries were also built around the cable car stations. In the intervening period the murder rate has halved.
- PlanIT Valley, near Paredes, Portugal: planned for 2015, this new eco-city will have a central computer acting like the city's brain (Knight, 2010). The computer will use data collected from sensors akin to a nervous system to control the city's power generation, water use and waste treatment.

There are of course also options for low-technology interventions that bring practitioners together, and avoid 'discipline apartheid' through improved knowledge management. Take for instance the case of helping dysfunctional families in Croydon through early intervention on social care. What is most insightful about this is how the different key service providers are coming together to prevent a social problem and reduce the cost burden at the same time.

Early interventions on social care in Croydon (UK)

Addressing dysfunctional families' difficulties before they become unsolvable is not only good news for the individuals, and the wider community, but can also save large amounts of money in the long term (Tickle, 2010).

According to Croydon Council the estimated current cost of families known to be users of multiple services is nearly US$100 million, for the group of two- to three-year-olds calculated up to the age of 18. If the early interventions identified by the council's department of children's services are successful the first 3 years would save US$8 million and after 5 years almost double to US$17 million.

The project was developed under a so-called 'total place' pilot whereby different public service agencies – social care, doctors and police amongst others – combined their efforts on themes of mutual concern.

Whilst interesting, however, the Croydon case does not explicitly involve the service users or other non-state actors in the identification of the leverage point, nor provide an oversight of how performance against these early interventions in social care are carried out, as recommended by Meadows (1997). The fascinating example of place-shaping in Michigan in the US does, however.

Place-based governance in Michigan (USA)

The state of Michigan awoke to the importance of 'place' in economic development over a decade ago (Project for Public Spaces, 2011). Key to this was recognising the importance of the quality of life, competing for the best talent, and attracting new and retaining existing businesses.

This so-called 'place-based' or 'place-shaping' strategy to economic development is about each element of community – residents, local government, private sector, schools and higher education, and NGOs – contributing to the overall success of its area by understanding, mapping and pooling intelligence and leveraging resources.

The Grand Traverse Bay Region in the state of Michigan is a long-term blueprint announced in 2011 to guide growth for the next half century and provide a vision for cooperation on transportation, education and broadband infrastructure to renew the area. This type of place-making is holistic and community-led, which means breaking down silos between agencies, building legitimacy to impacted stakeholders and providing a more accountable form of governance. A further benefit to Michigan during the austerity spending cuts is that the approach allows each tax dollar to do more.

Results here, and in similar place-based approaches, are so compelling that in 2011 UN-Habitat adopted the first ever public space resolution urging the development of a policy approach for the international application of place-making.

So, what we can decant from this insightful work on placed-based governance is that a better understanding and operation of a complex system – such as a collection of areas that make up a city – is obtained by bringing together a variety of actors with experiential knowledge, as well as the appropriate skill sets. Or another way to interpret this is that local resilience endeavours like this in Michigan make our societies more sustainable. And whilst these places are harder to make, they are certainly also harder to break.

So, moving from interventions and their frequency and efficiency, a new theory of resilience needs to emerge to address the challenges raised so far, and to draw together all the 'elements' of a resilient society.

Conclusions

- Localism without local government will work best through more strategic but less frequent interventions by municipal authorities.
- This will require unleashing the positive power of the markets and people alike.
- Smart interventions will require going beyond simply hoarding information to intelligently managing complex systems.
- Success will ultimately depend on how well local leaders use the key leverage points. Place-based governance is one powerful way to do this.

11

INFUSED RESILIENCE

A theory of change

Energy security equals national security equals quality of life.

Patrick Hays, Mayor, City of North Little Rock (Hays, 2011)

A refined interpretation of resilience

Numerous and diverse definitions of resilience abound (as we saw in Chapter 1) and the intention here is not to add to this list, rather to refine and develop how it is best delivered through the prism of sustainable urban development. As has been stated from the outset, local government is the key 'unit of currency' in building more sustainable societies.

In this regard, ICLEI–Local Governments for Sustainability (ICLEI, 2011a) usefully describe resilience as 'the ability of urban systems to withstand certain levels of stress', and that this resilience is achieved by: having flexible systems to absorb shocks; distributing stress across systems; restoring functionality in a timely way; having substitution options available; and building capacity to identify problems in advance in order to adapt. Whilst useful, however, like the other definitions and interpretations offered in Chapter 1, it is not a sufficient definition, as it does not entirely address the real world we live in – a world in which local government is part of a wider, complex system; a world in a new era of catastrophe ranging from tsunamis and floods to hurricanes and earthquakes that is simultaneously challenged to live within limited natural resource flows; a world in which new forms of PPPs have emerged to run our cities, towns and villages without complete agreement as to what an appropriate accountability regime ought to be.

Paramount to refining our approach to resiliency is an understanding of the shifting interface between citizen, state and other actors, which constitute the complex system of rules under which our societies are governed (depicted in Figure 11.1).

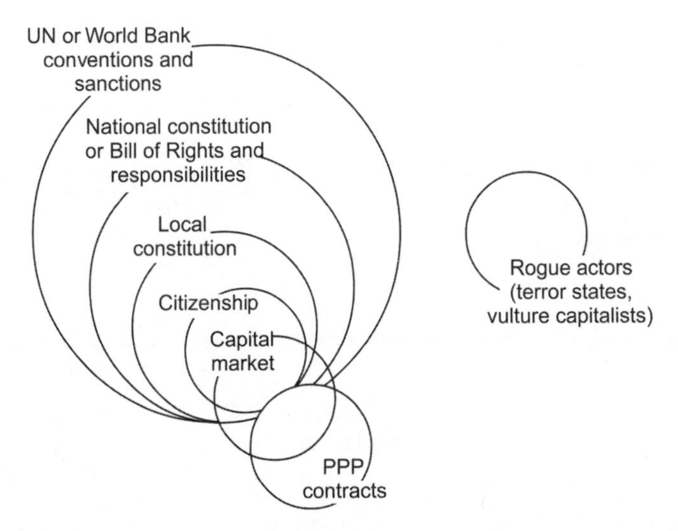

FIGURE 11.1 Governing sustainable urban development as a complex system (© Monaghan, 2012)

An appreciation from the outset of these synergies, overlaps and limitations or enablers will inform how we go about selecting the key leverage points, for instance how local constitutions do or do not work in harmony with national constitutions, or how PPP contracts are or are not accountable to the citizens they serve. Building on this interplay and through the learning about how to build more sustainable societies given the parlous state of our fragile planet, economy and communities, a new theory of change is proposed here – *infused resilience* – whereby power is dispersed intelligently as an enabler to positive and lasting change, as shown in Figure 11.2.

Each of these five stages or system conditions is mutually dependent and so must be observed to effect positive change. This may warrant the local authority reassigning staff to a dedicated new office, for instance the post of Chief Information Officer might become the 'Chief Resilience Officer', whose responsibilities would relate to ensuring key leverage points in the council's complex system across a host of departments and disciplines are fully utilised in support of its business strategy. The work of this resiliency office should be overseen by a scrutiny committee of elected members and local partners to ensure it has real power, supports place-based governance and delivers big results.From these key characteristics and leverage points a set of performance metrics can be determined, and a diagnostic applied to gauge how well the transition to resiliency is (or is not) going, as shown in Figure 11.3.

This dummy model here compares imaginary cities 'Mal' and 'Infrangilis' (mal being the Latin word for bad and infrangilis Latin for unbreakable). Regardless of whether 'Mal' outperforms other cities in other regards (such as GDP or food security) if it fails to meet all of the five quality thresholds or benchmarks set by Infrangilis then it is likely to falter, whether due to a failure to decouple vested interests from governance, the homogeneity of its financial service providers or a lack of scrutiny of PPPs. Further information about this diagnostic is available from the author's website www.infrangilis.org

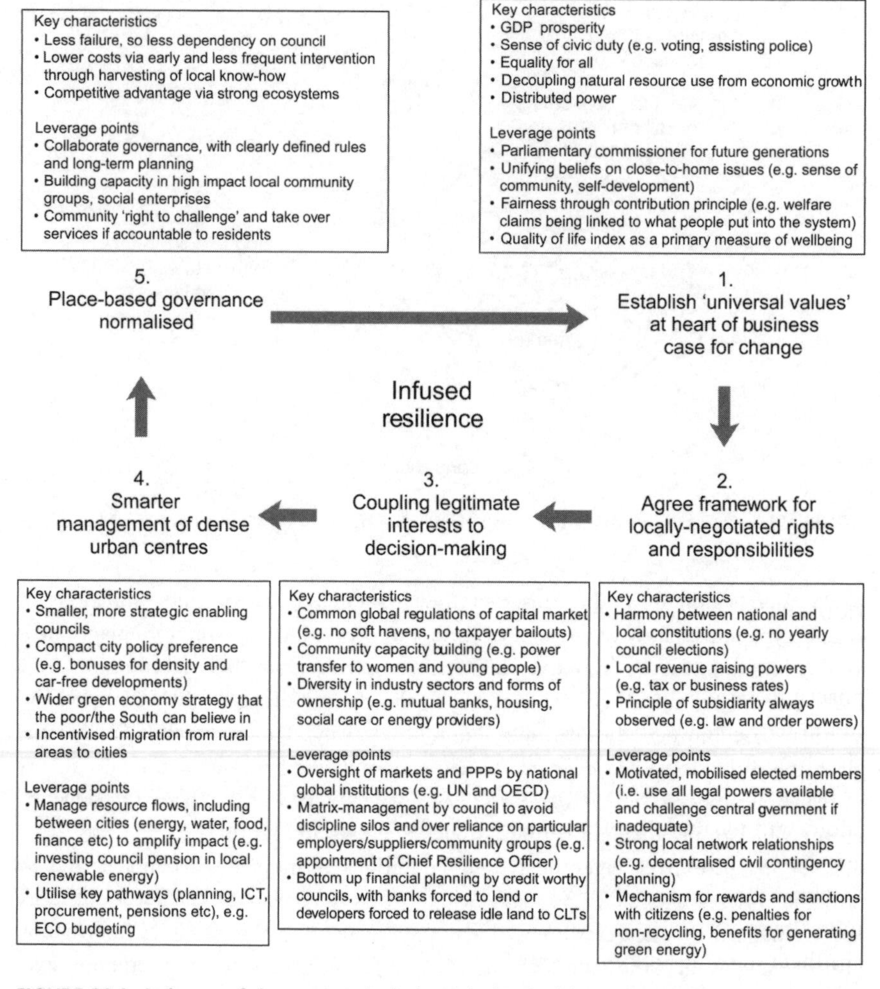

FIGURE 11.2 A theory of change to unlock the right kind of dependency (© Monaghan, 2012)

In saying this, of course, each city's context will differ from region to region, according to country politics, financial well-being and so on. National government, the business community and ordinary people will all have a vital role to play too. As such, the dilemma for city leaders is doing the very best they can, given their particular situation. This includes challenging any status quo, nationally as well as locally.

Embedding and maintaining empowerment

Whilst we know the current economic model is fundamentally flawed in the way it discounts the high cost of failure, and that prevention of this can save a lot of money (e.g. banking collapse, violence against women, droughts), let us also be very clear that to

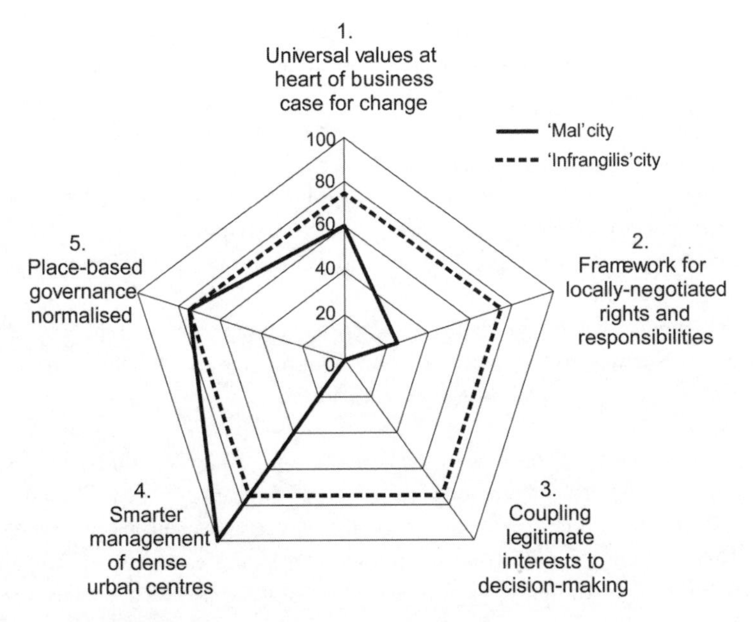

FIGURE 11.3 Infused resilience diagnostic (© Monaghan, 2012)

finance the transition towards more resilient societies, the costs are quite intimidating. Just in terms of climate adaptation, it is estimated that urban areas will be expected to absorb up to 80 per cent of the US$80–100 billion costs per annum (ICLEI, 2011b).

Let us also be equally blunt: lack of money is not the problem. We can, after all, afford warmongering year after year (Brown, 2011). Rather, it is an *unwillingness* to act, due to either selfishness or ignorance about the art of the possible.

This will mean local leaders sweating more money from the system by becoming more 'business savvy' in two key ways. Firstly, in how they overcome 'discipline apartheid' to craft bottom-up 'investment brochures' to be pitched to national government or investment houses or enter into new private sector arrangements, whether it be insurance cover, bank loans or PPP. This in turn should be complemented by banks being forced to lend.

The broader concept of the new 'business savvy' council is explored in more detail in Box 11.1.

Box 11.1 The brilliant council of the future?

According to global accountancy firm KPMG and The Centre for Public Service Partnerships (2011), to survive and prosper in an unprecedented age of austerity 'brilliant councils of the future' will need to:

- have highly skilled political leaders, high calibre councillors and a cabinet focused on outcomes and financial control;

- have a more accountable Chief Executive and council management team;
- frame a commercial business plan for the council and possibly deploy a new operating system to implement it;
- reduce the behavioural autonomy of directorate silos within the council, perhaps through matrix management;
- deploy an 'iron-like' grip on finances, whereby local citizens, like shareholders, get maximum returns for their local tax contributions;
- markedly increase the productivity and performance of the council's staff and suppliers;
- implement payment by results for anything it procures in its supply chain;
- enthusiastically cede control to local communities, as a critical tactic to secure better results for customers and citizens.

However, whilst another welcome contribution, caution should be exercised against the report's recommendation that a 'brilliant council' should be 'agnostic' about which part of the economy delivers service provision given, as we have seen in Chapters 8 and 9, the right checks and balances need to be in place if this involves for-profit providers for critical services such as social care.

Secondly, and perhaps more importantly, an increasingly business-like approach is also about how local assets can be best mobilised by leaders – in particular, people, planning, procurement and pensions. Returning to climate challenge as an illustration, these actions may range from devolving responsibility for civil emergencies to residents, controlling development through compact planning regimes (as outlined in Chapters 4 and 5 respectively), levying a flood tax on high-earning citizens (BBC, 2011b) and/or investing local government pension funds in decarbonised regeneration schemes (APSE, 2010). Which, in turn, need to be enabled through national policy setting. By way of illustration, the potential use of council pension funds is explained in more detail in Box 11.2.

Box 11.2 Investing local government pension funds in decarbonisation regeneration schemes

According to lawyers Walker Morris a council can, and should, use local authority pension funds to invest in local regeneration projects if it does so in a responsible way (APSE, 2010). This is a good way of maximising the use of existing resources and getting value for money in an age of financial austerity.

Taking the UK as an example, the Government Pension Scheme (Management and Investment of Funds Regulations, 2009) does not allow local councils to fund their expenditure from their pension funds (requiring it to be kept instead in a separate dedicated bank account from 1 April 2011). The administering authority for those funds still invests in regeneration projects (provided it is within the Regulations). To do so the pension funds' Statement of Investment Principles (SIP) needs to be broad enough to allow for the investment, and the associated risk profile. To reduce risk, councils could jointly invest with other surrounding local authorities to reduce any risk and benefit from economies of scale from larger schemes.

So, given that these pension funds are constituted from local authority officer contributions (many of whom will reside in the area in which the council operates), and from council funds (which are ultimately made up from revenue generated from local taxpayers), surely it is only right that this money is put to good use locally? Especially so if its funds are invested in a low-carbon economy scheme that creates local wealth in an environmentally and socially progressive way?

To realise the objective of *infused resilience* will, of course, require having the right type of leadership. It will, of course, necessitate everyone else playing their part too – this includes you (the next steps for each of us are in the final chapter).

Conclusions

- We need to refine the way we interpret resilience. At its core should be the new concept of 'infusion', that is, intelligent dispersal of power.
- This means local and national partners working collaboratively to establish common beliefs, agree rights and responsibilities, encourage a more compact way of living, decouple vested interests and normalise place-based governance.
- Ultimately, it is about doing the very best you can given your particular context, which may include challenging the status quo nationally as well as locally.
- Five stages need to be observed for societies to work effectively and, ultimately, become resilient: universal values at the heart of the business case for change, a framework for locally negotiated rights and responsibilities, coupling legitimate interest to decision-making, smarter management of dense urban centres and place-based governance becoming normalised.
- Associated costs may be high, but so are the savings, and this is a choice.
- Discipline apartheid (silo working across council departments) must end in order for this to succeed. A resilience office should be set up with oversight through a scrutiny committee of elected members and other key local partners.

12

WHAT YOU NEED TO DO NEXT

> The environment and poverty are twins and ignorance is their parents.
>
> Didas Massaburi, Mayor, Dar es Salaam (Massaburi, 2011)

The right type of local leadership

You will recall that this book began with a number of high-level questions:

- Why, when it comes to sustainable urban development, has 'business as usual' failed?
- Why do national and local leaders need to move away from this?
- What action can municipal authorities and community champions take to create more sustainable societies?
- How can they do this as the world attempts to recover from an age of austerity?
- How can multilateral agencies such as the UN or OECD support the transition to a green economy through better governance arrangements?
- How should the role of the private sector in new major urban infrastructure be scrutinised?

Hopefully you will agree that big answers to these questions have been proposed. But what now for our leaders who heed this call to arms?

First and foremost, leaders in local government need to step up to the challenge and be courageous. This will certainly require a new breed of 'positive deviant' as Parkin (2010) affectionately calls them. People who are intelligent and brave enough not to always toe the line or be force fed top-down solutions by politicians or experts, but rather understand that what the public values is determined with wider stakeholders.

More than this, though, new types of leaders who understand and work across a number of disciplines to craft the compelling narrative for taking action on difficult issues in order to best serve the public (Monaghan, 2010) are also essential. An impressive example of this is the commitment of a group of city mayors to the climate cause despite the inertia of their own national leaders, as per Box 12.1.

Box 12.1 World mayors sign climate change pact

In late 2010, 138 cities pledged their commitment to climate action as part of the Mexico City pact (World Mayors Council on Climate Change, 2010). Signatories included, amongst others, Barcelona, Buenos Aires, Istanbul, Johannesburg, Kyoto, Los Angeles, Paris, Rio de Janeiro and Sao Paolo. The Pact was timed to coincide with the COP16, the UN Conference on Climate Change.

A key component of the pact is the carbon cities registry – cCCR – which is intended to support the global credibility of local climate action by allowing transparency and comparability of climate commitments and performance.

Importantly, whilst national leaders failed, yet again, at COP16, to deliver a grand single deal on climate action it is inspiring that local councils are stepping in to fill this leadership void.

But, as you will know from reading this book, the buck does not just stop with local government leaders – we all have a crucial role to play. In particular, multilateral agencies, national policy makers, business groups and communities all have important (and distinct) responsibilities to enable the co-production of local solutions.

Box 12.2 Partner contributions to delivering infused resilience

Taking the *infused resilience* theory (see Chapter 11) as the key point of reference, Table 12.1 maps the contributions of each partner to delivering this, ranging from multilateral agencies such as the UN to people in local communities. (For national and local government, these are indicative examples, noting the context from region to region will vary).

Let us take one of the partner collaborations listed in Table 12.1 and elaborate. So, for instance, as a local resident, in respecting a 'sense of civic duty', you can do your bit by joining a local residents' association to help shape your local area. It may take the form of small things at first, such as challenging your neighbours if you know their children are a nuisance or if they do not recycle their household waste. You could contact your municipal authority to ask about the area's resiliency strategy and how you can contribute to it in order to ask questions about what they buy from local businesses, how they are dealing with local food and energy security, or how they are ensuring there is a diversity of local financial providers to households, small businesses and social enterprises. (If they send you the council's civil contingencies plan, which could include how they would deal with floods or terrorist attacks, return it saying that you requested something completely different.) In addition, you

TABLE 12.1 Partner contributions to delivering infused resilience

THEME	*PARTNER*				
	Multi lateral agencies	*National government*	*Local government*	*Business*	*Local communities*
1. Establish 'universal values' at heart of business case for change	–	Quality of life as primary measure of prosperity Appoint commissioner for future generations	Decouple natural resource use from growth strategy Promote fairness through contributory principle	Decouple natural resource use from growth strategy	Respect for sense of civic duty
2. Agree framework for locally-negotiated rights and skills	–	A constitution setting out rights *and* responsibilities Transfer revenue raising and policing powers to cities	Implement mechanism for local rewards and sanctions Nurture strong local network relationships	–	Support/scrutinise elected members' decision-making Participate in decentralised civil contingency plans
3. Coupling legitimate interests to decision-making	G20, IMF: redefine regulation of markets OECD, World Bank: create observatory for new PPPs	Lobby for global regulations of capital markets Ensure strong oversight of markets and PPPs	Nurture diversity in forms of ownership as well as industry Bottom-up financial planning Establish resiliency office	Contribute to development of common global regulations of markets and standards for PPPs	–
4. Smarter management of dense urban areas	A single, clear approach to interface between urbanism and green economy by UN family (i.e. Habitat, UNEP)	Incentivise migration from rural areas to cities Compactness policy preference	Smaller, more strategic enabling council Manage resource flows as a priority	Collaborate with councils to utilise key pathways (e.g. planning, procurement)	–
5. Place-based governance normalised	–	–	Establish place-based forums Build capacity in community groups and social enterprises	Participate in place-shaping forums	Participate in place-shaping forums Challenge malpractice or weak decision-making

could write to your locally elected politician and find out how your local constitution helps ensure the future prosperity for your family and community (through place-based governance), or even consider standing for election yourself to ensure other people voices' of today – and those of tomorrow – are heard too.

Key to you taking action, depending on who you are and what motivates you, is to always keep 'doing your bit' and asking questions about how others are also contributing in order to shake things up and make something happen:

- This is not to attack, but rather to support.
- You have the right to do so. Indeed, you have a responsibility to do so.

As the old adage goes – 'sunlight is the best disinfectant' (Louis Dembitz Brandeis, 1856–1941) – so let us direct some of those lovely, free rays of light on our toxic model. A discredited model which has routinely shattered lives, wrecked businesses and damaged the planet through a series of disasters ranging from water, food and energy scarcity to the global banking collapse. The time is right for a new model. This new model must have *infused resilience* at its heart.

Conclusions

- The current model is flawed in the way it discounts the high cost of social and environmental failure with the 2008/09 global banking collapse being just the latest example.
- *Infused resilience* is a new way of thinking about how best to create more sustainable societies that everyone can benefit from. At its core is the transfer of power from vested interests to ordinary people at the local level.
- Cities are at the heart of this solution, in terms of the better use of limited natural resources and finance and in advancing equality and social mobility.
- We can make the big changes needed to make if we really want to. But it will take courage as well as vision to make things happen.
- We all have a responsibility to challenge the status quo – from the UN to citizens themselves. What are you waiting for?

Appendix 1
AUTHOR'S BIOGRAPHY

As founder and CEO of Infrangilis, Philip is a recognised leader with over 17 years of international experience as a strategist and change manager in the fields of economic development and environmental sustainability. This includes half a decade with think tank AccountAbility, where he helped establish their global leadership on collaborative governance, responsible competitiveness and corporate responsibility standards.

An accomplished public speaker and writer on such matters, Philip provided expert opinion for television and newspapers including the BBC and the *Financial Times*. He is also the author of the acclaimed book *Sustainability in Austerity* (2010, Greenleaf Publishing) which was named as one of 'The 2010 Top 40 Sustainability Books' by the University of Cambridge Programme for Sustainability Leadership and which has been acclaimed by respected commentators from the UN, Harvard University, WWF and Accenture.

Philip works as an adviser on a variety of consultancy and research assignments across the public, private, civil society and academic sectors. Partners and clients have included the UN, the Albert Luthuli Centre for Responsible Leadership (at the University of Pretoria), Consumers International, the European Commission, the UK Department for Business, Innovation and Skills (BIS), North East Lincolnshire Council, Knowsley Metropolitan Borough Council, BP, Gap Inc, KPMG, Marks and Spencer, and Nike.

He is also an elected Fellow of the Royal Society for the Encouragement of Arts, Manufactures and Commerce (RSA), guest lecturer at the University of Liverpool's School of Environmental Sciences and a reviewer for the *Sustainability Accounting, Management and Policy Journal* edited by La Trobe University in Australia and published by Emerald.

Philip's previous publications include:

Monaghan, P. (2010) *Sustainability in Austerity: How Local Government Can Deliver During Times of Crisis* (Sheffield: Greenleaf)

Monaghan, P., Business for Social Responsibility, Weiser, J. (2003) *Business and Economic Development: The Impact of Corporate Responsibility Standards and Practices* (London: AccountAbility; San Francisco: Business for Social Responsibility; Branford, CT: Brody Weiser Burns)

Rubbens, C., Monaghan, P., Bonfigioli, E. and Zadek, S. (2002) *Impacts of Reporting: Brussels: The Role of Social and Sustainability Reporting in Organisational Transformation* (Brussels: CSR Europe; London: AccountAbility)

Philip welcomes shared learning and connections at www.infrangilis.org.

Appendix 2

OTHER HELPFUL SOURCES OF LEARNING

Local government networks

Association for Public Service Excellence (APSE) is a not-for-profit company specialising in local authority front line services ranging from environmental services to housing. www.apse.org.uk

CEOs for Cities is a civic lab of urban leaders catalysing a movement to advance the next generation of great American cities with a particular focus on talent, connections, innovation and distinctiveness. www.ceosforcities.org

Commonwealth Local Government Forum works to promote and strengthen effective democratic local government throughout the Commonwealth and to facilitate the exchange of good practice in local government structures and services. www.clgf.org.uk

Covenant of Mayors is a commitment by signatory towns and cities to go beyond the objectives of EU energy policy in terms of reduction in CO_2 emissions. www.eumayors.eu

C40 Cities is a group of large cities committed to tackling climate change through effective partnership working with the Clinton Climate Initiative. www.c40cities.org

ICLEI–Local Governments for Sustainability is an international association of local governments and their associations that have made a commitment to sustainable development. www.iclei.org/

The Local Government Information Unit (LGiU) is an award-winning think tank and membership organisation whose mission is to strengthen local democracy to put citizens in control of their own lives, communities and local services. https://member.lgiu.org.uk/

The National League of Cities is dedicated to helping US city leaders build better communities by serving as a resource to and an advocate for its membership of 1,600 municipalities. www.nlc.org/

Unite Cities and Local Governments (UCLG) represents and defends the interests of local governments on the world stage, with a membership of over 1,000 cities across nearly 100 countries. www.cities-localgovernments.org/

World Mayors Council on Climate Change is an alliance of committed local government leaders advocating an enhanced recognition and involvement of Mayors in multilateral efforts addressing climate change. www.worldmayors council.org/

Intergovernmental organisations

Food and Agriculture Organization (FAO) leads international efforts to defeat hunger for the UN. www.fao.org

Intergovernmental Panel on Climate Change (IPCC) is the lead body for the assessment of climate change, established by the United Nations Environment Programme. www.ipcc.ch

International Energy Agency (IEA) is an intergovernmental organisation which acts as energy policy advisor to countries in their effort to ensure reliable, affordable and clean energy. www.iea.org

Organisation for Economic Co-operation and Development (OECD) promotes policies to improve the economic and social wellbeing around the world by providing a forum for governments to work together. www.oecd.org/

United Nations Development Programme (UNDP) aims to build effective and capable states that are accountable and transparent, inclusive and responsive. www.undp.org

UN Educational, Scientific and Cultural Organization (UNESCO) works to create the conditions for dialogue among civilisations, cultures and peoples, based upon respect for commonly shared values. www.unesco.org.uk/

United Nations Environment Programme (UNEP) provides leadership and encourages partnership in caring for the environment by inspiring, informing and enabling nations and people to improve their quality of life. www.unep.org

UN-Habitat The United Nations Human Settlements Programme is the UN agency responsible for human settlements. It is mandated by the UN General Assembly to promote socially and environmentally sustainable towns and cities with the goal of providing adequate shelter for all. www.unhabitat.org

Non-governmental organisations and pressure groups

Avaaz (meaning 'voice' in several European, Middle Eastern and Asian languages) aims to organise citizens of all nations through the Internet to close the gap between the world we have and the world most people everywhere want. www.avaaz.org

Foundation for Democracy and Sustainable Development is a charity working to develop ideas and innovative practices so that democratic decision-making can work better for sustainable development. www.fdsd.org

Forum for the Future is an independent non-profit organisation with a mission to promote sustainable development. www.forumforthefuture.org

Governance International is a non-profit organisation which helps agencies to achieve outcomes for citizens through excellence in public governance. www.govint.org

Greenpeace is a global campaigning organisation that seeks to protect the natural environment and promote peace. www.greenpeace.org

The New Economy Working Group aims to contribute to reframing the economic policy debate to address the social and environmental imperatives and opportunities of the twenty-first century. www.neweconomyworkinggroup.org

Northern Alliance for Sustainability (ANPED) is an international not-for-profit organisation with a mission to empower Northern civil society through capacity development, exchanges and sharing of knowledge. www.anped.org

Sustainable Cities International (ICSC) is a partnership between three levels of government, the private sector and civil society organisations to catalyse action on urban sustainability in cities around the world. sustainablecities.net

Transparency International is the global civil society organisation whose mission is to create change towards a world free of corruption by raising awareness and diminishing apathy and tolerance of corruption, and devising and implementing practical actions to address it. www.transparency.org

The World Wildlife Fund (WWF)'s mission is to stop the degradation of the planet's natural environment and to build a future in which humans live in harmony with nature. wwf.panda.org

Academic, research or professional institutions

The African Centre for Cities is an interdisciplinary research and teaching programme focused on quality scholarship regarding the dynamics of unsustainable urbanisation processes in Africa, with an eye on identifying systemic responses. africancentreforcities.net

The Centre for International Governance Innovation (CIGI) is an independent, non-partisan think tank on international governance. www.cigionline.org

The Centre for Local Economic Strategies (CLES) is an independent thinking–doing organisation, with charitable status, involved in regeneration, local economic development and local governance. www.cles.org.uk

The Earth Policy Institute is an international think tank providing a vision of an environmentally sustainable economy. www.earthpolicy.org

Infrangilis is part think tank and part consultancy working to instigate or accelerate innovative solutions on the interface between the green economy and sustainable urban development. www.infrangilis.org

The Institute of Environmental Management and Assessment (IEMA) promotes best practice standards in environmental management, auditing and assessment. www.iema.net

LSE Cities is an international centre at the London School of Economics and Political Science (LSE) that carries out research to study how people and cities interact in a rapidly urbanising world, focusing on how the design of cities impacts on society, culture and the environment. www2.lse.ac.uk/LSECities/home.aspx

The New Economics Foundation (NEF) is an independent think and do tank that inspires and demonstrates real economic wellbeing. www.neweconomics.org

The Royal Society for the Encouragement of Art, Manufactures and Commerce (RSA) seeks to develop and promote new ways of thinking about human fulfilment and social progress. www.rsa.org

Urban Labour Network is a global knowledge-sharing network instigated by the International Labour Organization (ILO) with a specific focus on the role of labour in urban development. www.urban-labour.net

SURF is a dedicated centre within the University of Salford, established to undertake interdisciplinary research on Sustainable Urban and Regional Futures (SURF). www.surf.salford.ac.uk

Sustainability Institute is a non-profit organisation which applies systems thinking, system dynamics modelling and organisational learning to economic, environmental and social challenges. www.sustainabilityinstitute.org/

World Resource Institute is an environmental think tank that goes beyond research to find practical ways to protect the earth and improve people's lives. www.wri.org

Business networks

AccountAbility is a global organisation that works to promote accountability innovations for sustainable development. www.accountability21.net

Business for Social Responsibility is a global non-profit organisation whose mission is to work with businesses to create a just and sustainable world. www.bsr.org

CSR Europe is a network for businesses to share best practice on corporate social responsibility. www.csreurope.org

University of Cambridge Programme for Sustainability Leadership works with leaders from business to address the critical global challenges that affect the success of their organisations. www.cpsl.cam.ac.uk/

US Green Building Council is a non-profit community of leaders working to make green buildings available to everyone within a generation. www.usgbc.org

World Business Council on Sustainable Development (WBCSD) aims to provide business leadership as a catalyst for change toward sustainable development. www.wbcsd.org

BIBLIOGRAPHY

Adger, N. (2010) 'An Interview with Neil Adger: Resilience, Adaptability, Localisation and Transition', http://transitionculture.org/2010/03/26/an-interview-with-neil-adger-resilience-adaptability-localisation-and-transition/ (accessed 10 October 2011)

Aldersgate Group (2006) *Green Foundations: Better Regulation and a Healthy Environment* (London: Aldersgate Group)

Amsler, T. (2010) 'Involve Youth in Local Planning', *Western City*, http://www.westerncity.com/Western-City/September-2010/Involving-Youth-in-Local-Planning (accessed 8 October 2010)

APSE (2010) 'Investing Local Government Pension Funds in Regeneration Schemes', *APSE direct news* (August/September: 12)

Arup (2010) *Smart Cities: Transforming the 21st Century City via the Creative Use of Technology* (London: Arup)

Associated Press (2011) 'California Sets Nation's Highest Renewable Power Goals', http://losangeles.cbslocal.com/2011/04/12/california-sets-nations-highest-renewable-power-goals/ (accessed 26 April 2011)

Banyard, K. (2010) *The Equality Illusion: The Truth About Women and Men Today* (London: Faber and Faber)

Barley, S. (2010) 'Escape to the City', *New Scientist* (6 November: 32)

Bawden, A. (2010) 'New Frontiers', *Guardian* (29 September: 1)

BBC (2010) 'India and China set $100bn Trade Target by 2015', http://www.bbc.co.uk/news/world-south-asia-12006092 (accessed 14 July 2011)

——(2011a) *Filthy Cities: Revolutionary Paris* (London: BBC)

——(2011b) 'Australia floods: PM Julia Gillard Unveils New Tax', http://www.bbc.co.uk/news/world-asia-pacific-12294834 (accessed 27 January 2011)

Beament, E. (2011), 'Rise in Climate Change Disasters', *Independent*, http://www.independent.co.uk/environment/climate-change/rise-in-climate-change-disasters-2281160.html (accessed 9 May 2011)

Benello, C. G., Swann, R. and Turnbull, S. (1997) *Building Sustainable Communities: Tools and Concepts for Self-Reliant Economic Change* (New York: The Bootstrap Press)

Black, I. (2011) 'Al-Qaida's Influence Slips from Marginal to almost Irrelevant in Light of Arab Spring', *Guardian* (3 May: 16)

Bollier, D. (1998) *How Smart Growth Can Stop Sprawl* (Washington, DC: Essential Books)

Bowers, S. (2011) 'Lost Your Big Bonus? Have a Pay Rise Instead', *Guardian* (26 April: 1)

Branigan, T. (2011) ‘No Promotion for Chinese Officials Who Neglect Family’, *Guardian* (22 January: 22)
Bresch, D. N. and Spiegel, A. (2011) *Economics of Climate Adaptation: Shaping Climate-Resilient Development: A Framework for Decision-Making* (Zurich: Swiss Re)
Brooks, D. (2011) *The Social Animal: A Story of How Success Happens* (London: Short Books)
Brown, L. R. (2011) *World on the Edge: How to Prevent Environmental and Economic Collapse* (Washington, DC: Earth Policy Institute)
Brugmann, J. (2009) *Welcome to the Urban Revolution: How Cities are Changing the World* (Noida: HarperCollins)
Brunt, L. (2010) ‘It Was the Greatest Promise Ever Made: But Can We Keep it?’, *Guardian* (20 September: 17)
Bunting, M. (2011) ‘Outrage at the Banks is Everywhere, so Why aren’t there Riots on the Streets?’, *Guardian* (30 May: 23)
Burke, J. (2010) ‘A Community Approach to Justice’, *Guardian* (14 November: 11)
Butt, R. (2011) ‘Archbishop of Canterbury Says Rich Should Help Poor’, http://www.guardian.co.uk/uk/2011/apr/21/archbishop-canterbury-rich-help-poor (accessed 18 May 2011)
Carrell, S., Wintour, P., Gumbel, A. and Dodd, V. (2011) ‘Cameron Looks for Lessons from LA to Tackle Homegrown Gangs’, *Guardian* (12 August: 6)
Carroll, R. (2010) ‘Developed Countries’ Carbon Cancelled Out by Exports’, *Guardian* (15 September 2010: 19)
Cartwright, A., Parnell, S., Oelofse, G., and Ward, S. (forthcoming May 2012), *Climate Change at the City Scale: Impacts, Mitigation and Adaptation in Cape Town* (London and New York: Routledge).
Castells, M. (1990) *The Information City: A Framework for Social Change* (Toronto: University of Toronto)
CEOs for Cities (2011) *Compact Cities Save Residents Money in the Face of Rising Gas Prices*, http://www.ceosforcities.org/blog/entry/3011 (accessed 30 March 2011)
Chanan, G. (2009) *Valuing Community Empowerment: Making the Business Case* (Taunton: The SW Regional Empowerment Partnership)
Chang, H. (2010) *23 Things They Don’t Tell You About Capitalism* (London: Allen Lane)
Changwon (2006) *Environmental Capital Changwon* (Changwon: Changwon City Government)
Christie, J. (2011) ‘Compelling Argument’, *Guardian* (12 January 2011: 1)
Clark, D. (2011) ‘Capitalist Storm Clouds Loom Over Havana After Sates Cuts 1m Jobs’, *Guardian* (26 April: 11)
CLES (2011) ‘Local Buying can Create Savings and Improve Communities’, *CLES News* (24 May)
Cohen, S. (1999) ‘A Conceptual Framework for Devolving Responsibility and Functions from Government to the Private Sector’, paper prepared for the 60th Annual Meeting of the American Society for Public Administration, Orlando (10–14 April)
Conaty, P., Birchall, J., Bendle, S., and Foggitt, R. (2003) *Common Ground for Mutual Home Ownership: Community Land Trusts and Shared-Equity Co-Operatives to Secure Permanently Affordable Homes for Key Workers* (London: New Economics Foundation/CDS Co-operatives)
Conrad, M. (2011) ‘City Group Calls for Rapid Merger of Local Budgets’, *The MJ* (26 May: 5)
Cooke, G. (2011) *National Salary Insurance: Reforming the Welfare State to Provide Real Protection* (London: IPPR)
Council of Global Unions (2009) *Getting the World to Work: Green Growth for Jobs and Social Justice* (Brussels: CGU)
Crompton, T. (2010) *Common Cause: The Case for Working with our Cultural Values* (Godalming: WWF)
Curtis, S. and Ramesh, R. (2010) ‘Local Groups Get Right to Take Over Services in “Big Society” Bill’, *Guardian* (11 December: 22)
Davutoglu, A. (2011) ‘*We have been insulted and humiliated. But finally history is bringing us dignity*’, edited extract of a speech at the 6th Al-Jazeera forum in Doha (*Guardian*, 16 March: 33)

De Visser, J. (2005), *Developmental Local Government: A Case Study of South Africa* (Antwerp: Intersentia).
Dean, J. (2011) 'Pennsylvania's Green Economy', *The Green Economy*, http://www.thegreeneconomy.com/pennsylvanias-green-economy/ (accessed 10 October 2011)
DirectGov (2011), 'Local Government Elections', http://www.direct.gov.uk/emGovernment citizensandrights/UKgovernment/ (accessed 27 April 2011)
Dobbs, R., Smit, S., Remes, J., Manyika, J., Roxburgh, C. and Restrepo, A. (2011) *Urban World: Mapping the Economic Power of Cities* (Seoul: McKinsey Global Institute)
Doward, J. and Stevens, J. (2011) 'Libraries, Youth Clubs and Street Cleaning Stand By to "Go Private"', *Observer* (16 January: 25)
Dunbar, R. (2011) 'Friends to Count on', *Guardian* (25 April: 12)
Economist, The (2010) 'The World in 2011', *The Economist* (22 November)
——(2011) 'Where do you Live? Town- and Country-dwellers have Radically Different Prospects', *The Economist* (23 June), http://www.economist.com/node/18832092?story_id=18832092&fsrc=rss (accessed 29 June 2011)
Elliott, L. (2010) 'Can Shopaholic Britain be Happy with Less?', *Guardian* (27 December: 27)
——(2011a) 'Could Fish Hold the Key to Financial Stability?', *Guardian* (18 April: 22)
——(2011b) 'Three Years On, It's as if the Crisis Never Happened', *Guardian* (30 May: 22)
——and Tran, M. (2010) 'Climate Change Could Derail Years of Progress in Improving Lives of World's Poor, says UN', *Guardian* (5 November: 28)
ENDS (2010) 'Cancún Progress Disguises Post-Kyoto Tensions' *ENDS Report* (December: 431: 6)
Environmentalist, The (2011a) 'Biodiversity Concerns', *The Environmentalist* (May: 8)
——(2011b) 'Taking Responsibility', *The Environmentalist* (May: 3)
FAO (2010) *Climate Smart Agriculture: Policies, Practices and Financing for Food Security, Adaptation and Migration* (Rome: Food and Agriculture Organization of the United Nations)
Forrester, J. W. (1971) *World Dynamics* (Portland, OR: Productivity Press)
Forum for the Future (2010) *The Sustainable Cities Index: Ranking the 20 Largest British Cities* (London: Forum for the Future)
Furedi, F. (2004) 'Heroes of the Hour', *New Scientist*, http://www.newscientist.com/article/mg18224463-heroes-of-the-hour.html (accessed 12 May 2011)
Garton Ash, T. (2011) 'The Optimists of Davos Past Now Face a World whose Script has Gone Awry', *Guardian* (27 January: 33)
Glaeser, E. (2011) 'Urban Ingenuity', *RSA Journal* (Summer: 16)
Goldenberg, J. (2011) 'Too Little, too Late: UN Chief Rejects Summits and Seeks New Way to Beat Global Warming', *Guardian* (28 January: 31)
Goodley, S. (2011) 'EU Inquiry into Bank Collusion Claims', *Guardian* (30 April: 40)
Goodyear, S. (2010) 'Is There a War between Cities and Suburbs? Does There Have to Be One?', *Grist*, http://www.grist.org/article/2010-10-20-is-there-a-war-between-cities-and-suburbs (accessed 7 June 2011)
Göpel, M. (2011) 'International Sustainability Think Tank Releases Draft "Principles for the Green Economy"', http://www.stakeholderforum.org/sf/outreach/index.php/intersesh2-item3 (accessed 11 January 2011)
Governance International (2011) *Public Service Co-Production: A Governance International Briefing Note* (Birmingham: Governance International)
Grayling, A. C. (2007) *Towards the Light* (London: Bloomsbury)
Green Alliance (2010) *Future Proof: An Electricity Network for the 21st Century* (London: Green Alliance)
Green Economy, The (2011) '$4.8 Million: Market Traction for Recyclers in Latin America', *The Green Economy*, http://www.thegreeneconomy.com/4–8-million-to-get-market-traction-for-recyclers-in-latin-america/ (accessed 7 June 2011)
Green Futures (2011) 'Bang to Environmental Rights', Forum for the Future, London (April: 6)
Guagnano, G., Stern, P. and Dietz, T. (1995) 'Influences on Attitude Behaviour Relationships: A Natural Experiment with Curbside Recycling'. *Environment and Behaviour*, 27, 699–718

Haldane, A. G. (2009) 'Rethinking the Financial Network', speech delivered at the Financial Student Association, Amsterdam, April

Hargroves, K. C. and Smith, M. H. (2005) *The Natural Advantage of Nations* (London: Earthscan)

Harvey, F. (2011) 'UK Left Behind in Race to Invest in Green Economy, Says Report', *Guardian* (29 March: 15)

Hawkes, A. and Wachman, R. (2011) 'Banks May Be Left to Collapse Next Time, says Moody's', *Guardian* (25 May: 28)

Hays, P. (2011) *Speaking at Resilient Cities 2011: 2nd World Congress on Cities and Adaptation to Climate Change* (Bonn, 3–5 June)

Heinburg, R. (2007) *Peak Everything: Waking Up to Decline in Earth's Resources* (East Sussex: Clairview Books)

Heine, J. (2011) 'Globalization and Democracy', Centre for International Governance Innovation, http://www.cigionline.org/articles/2011/05/globalization-and-democracy (accessed 30 June 2011)

Hickman, L. (2011) 'The Small Town that Came Up with Big Incentives to Recycle', *Guardian* (18 March: 26)

Hill, A. (2011) 'UK Uncut Turns Ire on RBS as Protests Spread to US', *Guardian* (28 February: 10)

HM Government (2011) *Enabling the Transition to a Green Economy: Government and Business Working Together* (London: HM Government)

Holling, C. S. (1973) 'Resilience and Stability of Ecological Systems', *Annual Review of Ecology and Systematics*, 4:1–23

Home Depot Foundation (2008) 'Case Study: Troy Gardens, Madison Area Community Land', http://www.homedepotfoundation.org/pdfs/aoe_home_madison.pdf (accessed 25 June 2011)

Hutton, W. (2010) *Them and Us: Changing Britain – Why We Need a Fairer Society* (London: Little Brown)

ICLEI (2008) *Amsterdam Food Strategy, the Netherlands: Urban–rural Linkages Enhancing European Territorial Competitiveness – Mini Case Study on Food Chains* (Freiburg: ICLEI–Local Governments for Sustainability)

——(2011a) *Urban Resilience: An ICLEI Briefing Sheet* (Freiburg: ICLEI–Local Governments for Sustainability)

——(2011b) *Financing the Resilient City: An ICLEI White Paper* (Freiburg: ICLEI–Local Governments for Sustainability)

——and UNESCO-IHE (2011) *Adapting Urban Water Systems to Climate Change: A Handbook for Decision Makers at the Local Level* (Freiburg: ICLEI–Local Governments for Sustainability)

ICSC (2009) *Centering Women in Reconstruction and Governance: 2006–2009 Sri Lanka Final Report* (Vancouver: ICSC)

Inman, P. (2011) 'UN Report Calls for Regulation to Curb Speculators Pushing Up Food Prices', *Guardian* (6 June: 20)

IEA (2009) 'Prospect of Limiting the Global Increase in Temperature to 2°C Getting Bleaker', International Energy Agency, http://www.iea.org/index_info.asp?id=1959 (accessed 22 July 2011)

International Resource Panel (2011) *Decoupling: Natural Resource Use and Environmental Impacts from Economic Growth* (Paris: UNEP)

Jenkins, S. (2010) 'Eric Pickles is Merely Hazel Blears in Super-size Wolf's Clothing', *Guardian* (15 December: 31)

Johnson, S. (2010) *Where Good Ideas Come From* (London: Allen Lane)

Jowit, J. (2010) 'WWF: Humans Use 1.5 Planets' Worth of Resources', *Guardian* (14 October: 15)

Kaplan, M. and Simpson, A. (2011) 'Round, Round Get Around', *Green Futures* (January: 18)

Keivani, R., Tah, J. H. M, Kurul, E. and Abanda, H. (2010) *Green Jobs Creation Through Sustainable Refurbishment in the Developing Countries* (Geneva: ILO)

Kingsley, P. (2011) 'The Voice of Protest', *Guardian* (28 February: 5)

Kintisch, E. (2010) *Hack to the Planet: Science's Best Hope – or Worst Nightmare – for Averting Climate Catastrophe* (Cranford, NJ: Wiley)

Knight, H. (2010) 'The Green City that has a Brain', *New Scientist* (6 October: 2781)

Korten, D. (2011) 'Mapping Unchartered Waters: A Compelling New Economy Policy Framework is being Framed and Developed', *CSRwire*, http://csrwiretalkback.tumblr.com/post/6497726746/mapping-uncharted-waters (accessed 14 June 2011)

KPMG and Centre for Public Service Partnerships (2011) *The Brilliant Local Authority* (London: KPMG and Centre for Public Service Partnerships)

Kunstler, J. H. (2005) *The Long Emergency: Surviving the Converging Catastrophes of the 21st Century* (London: Atlantic Books)

Lawrence, F. (2011) 'Barclays Faces Protests Over Role in Food Crisis', *Guardian* (26 April: 24)

LGC (2011) *The Most Influential Voices in Local Government* (London: Local Government Chronicle)

Liverpool Daily Post (2010) 'Pupils at St Aidan's Primary Grill Drivers Caught Using Mobile Phones', *Liverpool Daily Post*, http://www.liverpooldailypost.co.uk/liverpool-news/regional-news/2010/11/05/pupils (accessed 5 November 2010)

LSE and Deutsch Bank Alfred Herrhausen Society (2011) *Living in the Endless City* (London: Phaidon)

Lynas, M. (2011) *The God Species: How the Planet Can Survive the Age of Humans* (London: Fourth Estate)

McInroy, N. (2011) 'Stewardship of Place', *apse direct news* (May/June: 6)

——and Longlands, S. (2010) *Productive Local Economies: Creating Resilient Places* (Manchester: CLES)

McKinsey and Co. (2009) *Public–Private Partnerships: Harnessing the Private Sector's Unique Ability to Enhance Social Impact* (New York: McKinsey and Company)

McNeil, C. and Thomas, H. (2011) *Green Expectations: Lessons from the US Green Jobs Market* (London: IPPR)

Maplecroft (2011) *Water Stress Index 2011* (Bath: Maplecroft)

Martinson, J. (2011) 'A New Superhero', *Guardian* (22 May: 16)

Massaburi, Didas (2011) Speaking at Resilient Cities 2011: 2nd World Congress on Cities and Adaptation to Climate Change (Bonn, 3–5 June)

Matsumoto, T. (2010) 'The Impacts and Challenges of Compact City Policies', presentation at TCPA Roundtable, London (16 November)

Mayo, E. and Tizard, J. (2010), 'A Mutual Feeling', *Guardian*, http://www.guardianpublic.co.uk/public-service-cooperatives (accessed 3 December 2010)

Meadows, D. (1997) *Leverage Points: Places to Intervene in a System* (Hartland, VT: Sustainability Institute)

Middleton, P. and Seddon, J. (2010) *Delivering Public Services That Work* (Devon: Triarchy Press)

Milne, S. (2011) 'The Forces Unleashed in Egypt Can't be Turned Back', *Guardian* (3 February: 35)

Ministry of Justice (2009) *Rights and Responsibilities: Developing Our Constitutional Framework* (London: HM Government)

Monaghan, P. (2010) *Sustainability in Austerity: How Local Government Can Deliver During Times of Crisis* (Sheffield: Greenleaf)

——(2011a) 'Case interview with Åsa Karlsson Björkmarker, Växjö' (13 September)

——(2011b) 'Case interview with Jane McRae and colleagues, ICSC' (20 May)

——(2011c) 'Case interview with Thurstan Crocket, Brighton and Hove City Council' (20 May)

——(2011d) 'Case interview with Anton Cartwright, African Centre for Cities' (27 May)

——(2011e) 'Case interview with Tadashi Matsumoto, OECD' (27 June)

——(2011f) 'Conversation with Mike Hodson, SURF' (11 March)

——(2011g) 'Case interview with Dr Iván Narváez, FLASCO', with special thanks to UNEP for assistance with all translations (4 October)

——(2011h) 'Conversation with Ed Mayo, Co-operatives UK' (23 March)
——(2011i) 'Case interview with Pat Conaty, New Economic Foundation' (18 April)
——(2011j) 'Case interview with Pim Vermeulen, City of Amsterdam' (30 May)
——Business for Social Responsibility and Weiser, J. (2003) *Business and Economic Development: The Impact of Corporate Responsibility Standards and Practices* (London: AccountAbility, Business for Social Responsibility and Brody Weiser Burns)
Moon, M. K. (2011) Speaking at Resilient Cities 2011: 2nd World Congress on Cities and Adaptation to Climate Change (Bonn, 3–5 June 2011)
Morris, K. (2010) 'Five Routes to the Future', *Guardian* (17 November: 7)
Moya, E. (2010) 'Over 100 American Cities Face Ruin in Third Wave of Global Debt Crisis', *Guardian* (21 December: 24)
Murray, K. (2011) 'High Anxiety as Cuts Bite Deep', *Guardian* (16 February: 6)
Narey, M. (2011) *We Need to See Progress Soon, And we Can Afford it, RSA Journal* (Spring: 29)
NEF (2008), 'Plugging the Leaks', New Economics Foundation, http://www.pluggingtheleaks.org/about/index.htm (accessed 30 June 2011)
——and Compass (2011) *Good Banking: Why we Need a Bigger Public Debate on Financial Reform* (London: New Economic Foundation)
New Economy Working Group (2011) *How to Liberate America from Wall Street Rule* (Washington, DC: New Economy Working Group)
Newman, P., Beatley, T. and Boyer, H. (2009) *Resilient Cities: Responding to Peak Oil and Climate Change* (Washington, DC: Island Press)
Norman, J. (2011) 'Patterns of Attachment', *RSA Journal* (Spring: 14)
OECD (2011) *Better Life Initiative: Better Life Index*, http://www.oecd.org/document/35/0,3746,en_2649_201185_47837411_1_1_1_1,00.html (accessed 25 May 2011)
——(2012a) *OECD Compact City Policies in the Toyama Metropolitan Region* (Paris: OECD)
——(2012b) *Compact City Policies: A Comparative Assessment* (Paris: OECD)
Ofgem (2010) *Low Carbon Networks Fund: Creating Britain's Low Carbon Future* (London: Ofgem)
O'Grady, S. (2011) 'Southern Cross Won't Be the Last Private "Partner" to Hit the Rocks', *Independent* (2 June: 5)
Ostrom, E. (1990) *Governing the Commons: The Evolution of Institutions of Collective Action* (New York: Cambridge University Press)
Parkin, S. (2010) *The Positive Deviant: Sustainability Leadership in a Perverse World* (London: Earthscan)
Phillips, L. (2011) 'Hungary Wants Mothers to be Given Additional Votes', *Guardian* (18 April: 1)
Poulter, S. (2011) 'It's Bananas: Fruit Gets a Second Skin with Del Monte Packaging', *Daily Mail*, http://www.dailymail.co.uk/news/article-1361666/Del-Monte-packaging-Bananas-second-skin.html (accessed 1 March 2011)
Prendergast, J. (2008) *Disconnected Citizens: Is Community Empowerment the Solution?* (London: SMF)
Project for Public Spaces (2011) 'Michigan Leads the Way', http://www.pps.org/blog/michigan-leads-the-way (accessed 13 May 2011)
Ramesh, R. (2010a) 'The Miserable Rich: Europe's Wellbeing Drops as Incomes Rise', *Guardian* (16 November: 9)
——(2010b) 'Poverty Not Race is at the Root of Community Mistrust, Says Study', *Guardian* (29 November: 18)
——(2010c) 'Road Death Toll Soars in Poorest Countries', *Guardian* (13 September: 16)
——(2011) 'Co-ops and *Txokos*: Why the Big Society Thrives in Basque Country', *Guardian* (30 March: 18)
Reed, A. M. and Reed, D. (2006) 'Corporate Social Responsibility, Public–Private Partnerships and Human Development: Towards a New Agenda (and Beyond)', paper presented at conference 'Public Private Partnerships in the Post-WSSD Context', Copenhagen 14–16 August

Roberts, D. (2011) 'Great Places: Smart Density as Part of Economic Flourishing', *Grist*, http://www.grist.org/cities/2011-05-31-great-places-smart-density-economic-flourishing (accessed 7 June 2011)

Roderick, P. (2010) *Taking the Longer View: UK Governance Options for a Finite Planet* (London: Foundation for Democracy and Sustainable Development and WWF)

——(2011) *The Feasibility of Environmental Limits Legislation* (Godalming: WWF-UK)

Rowson, J., Broome, S. and Jones, A. (2010) *Connected Communities: How Social Networks Power and Sustain the Big Society* (London: RSA)

Rubbens, C., Monaghan, P., Bonfigioli, E. and Zadek, S. (2002) *Impacts of Reporting: Brussels: The Role of Social and Sustainability Reporting in Organisational Transformation* (Brussels: CSR Europe; London: AccountAbility)

Rushe, D. (2011) 'Financial Myth-breaker Charts Tarnishing of a Golden Reputation', *Guardian* (29 April: 33)

Rustin, S. (2010) 'Missing Links', *Guardian* (1 December: 1)

Sarchet, P. (2011) 'New Technology Offer Benefits For All', *Guardian* (13 April: 3)

Saunders, D. (2010) *Arrival City: How the Largest Migration in History is Reshaping Our World* (Harlow: Heinemann)

Seva, V. S. (2011) Speaking at Resilient Cities 2011: 2nd World Congress on Cities and Adaptation to Climate Change (Bonn, 3–5 June 2011)

Shaw, K. (2010) 'The Rise of the Resilient Local Authority?', presentation to ICLEI-Local Governments for Sustainability agenda, Brussels (September)

Siddique, H. (2011) 'Iceland's Sons and dÓttirs crowdsource a new constitution', *Guardian* (10 June: 24)

Siegle, L. and Borden, H. (2011) 'Al-Qaida's Influence Slips from Marginal to almost Irrelevant in Light of Arab Spring', *Observer* (6 February: 28)

Smith, D. (2011) 'The Rail Way: How Lagos Hopes to End Daily Endurance Test and Change Lives of Millions', *Guardian* (15 January: 29)

Stanvliet, R. and Parnell, S. (2006) 'The Contribution of the UNESCO Biosphere Reserve Concept to Urban Resilience', *Management of Environmental Quality: An International Journal*, 17: 437–49

Stern, P. (2000) 'Towards a Coherent Theory of Environmentally Significant Behaviour', *Journal of Social Issues*, 56, 407–24

——(2005) 'Psychological Research and Sustainability Science', keynote address to the 6th Biennial Conference on Environmental Psychology, Bochum, Germany, 21 September

Stewart, H. and Loweth, K. (2011) 'Living Wage Campaign Marks 10 Years of Fighting for the Poorest', *Observer* (1 May: 40)

Stratton, A. (2010) 'Smile Please is the New GDP, Cameron Declares. But How to Measure it?', *Guardian* (15 November: 14)

Swilling, M., Robinson, B., Marvin, S. and Hodson, M. (2011) 'Growing Greener Cities', discussion paper commissioned by UN Habitat for Expert Group Meeting, Nairobi

Tibbetts, G. (2008) 'Brighton Aims to Become United Nations "Biosphere Reserve"', *Daily Telegraph*, http://www.telegraph.co.uk/earth/earthnews/3352867/Brighton-aims-to-become-United-Nations-biosphere-reserve.html (accessed 1 June 2011)

Tickle, L. (2010) 'Necessity of Intervention', *Guardian* (3 November: 1)

Townshend, T., Frankhauser, S., Matthews, A., Clément, F., Liu, J. and Narcisco, T. (2011) *GLOBE Climate Legislation Study* (London: GLOBE International and Grantham Research Institute on Climate Change and the Environment)

Toynbee, P. (2010) 'The Greatest Unhappiness for the Greatest Number', *Guardian* (16 November: 29)

——(2011) 'They Caused the Crash and Now they Strangle Recovery', *Guardian* (28 May: 39)

Treanor, J. (2011) 'US Warns of "Light Touch" Bank Rules that Damaged UK', *Guardian* (7 June: 23)

——and Elliott, L. (2011) 'Osborne Backs Plan to Ring-Fence Retail Banks', *Guardian* (6 June: 1)

UN DESA (2005) *Good Practices in Agricultural Water Management Case Studies from Farmers Worldwide* (New York: UN Department of Economic and Social Affairs)
UNEP (2009) *The Green Economy Initiative* (Paris: UN Environment Programme)
——(2011a) *IEA Training Manual: Volume Two: Themes: Climate Change Vulnerability and Impact Assessment in Cities: Final Draft* (Paris: UN Environment Programme)
——(2011b) *Towards a Green Economy: Pathways to Sustainable Development and Poverty Eradication – A Synthesis for Policy Makers* (Paris: UN Environment Programme)
UNESCO (2011) 'FAQ: Biosphere Reserves?', http://www.unesco.org.uk/uploads/biosphere%20reserves%20faq/pdf (accessed 1 June 2011)
UN-Habitat (2011) *What Does the Green Economy Mean for Sustainable Urban Development?* (Nairobi: UN-Habitat)
UN High-level Panel on Global Sustainability (2011) *Second Meeting of the Panel: Meeting Report* (Cape Town: 24–5 February)
UCLG (2007) *Decentralization and Local Democracy in the World* (Barcelona: United Cities and Local Governments)
University of Cambridge (2011) *A Journey of a Thousand Miles: The State of Sustainability Leadership 2011*, Cambridge Programme for Sustainability Leadership (Cambridge: University of Cambridge)
Vaccaro, I., Zanotti, L. C. and Sepez, J. (2009) 'Commons and Markets: Opportunities for Development of Local Sustainability', *Environmental Politics*, 18: 4, 522–38
Vaughan, A. (2010) 'After Zombies and Aliens, Computer Game Fans Face a New Challenge: Climate Change', *Guardian* (1 November: 3)
VicHealth (2007) *Preventing Violence Before it Occurs: A Frame Work and Background Paper to Guide the Primary Prevention of Violence Against Women in Victoria* (Victoria: State Government Victoria)
Vidal, J. (2011a) 'Mother Earth Earns Human Rights under New Bolivian Law', *Guardian* (11 April: 15)
——(2011b) 'Huge Mirrors and Aerosols in Space: Just Science Fiction or the Earth's Last Hope?', *Guardian* (16 June: 13)
Walker, P. (2011) 'WikiLeaks and Guardian Hailed as Catalysts of Arab Spring', *Guardian* (13 May: 17)
Ward, H. (2009) 'Duty to Involve', http://www.fdsd.org/tag/duty-to-involve/ (accessed 8 August 2011)
Watts, J. (2010) 'China Counts £130bn Cost of Economic Growth', *Guardian* (29 December: 10)
——(2011) 'Amid a Boom Built on Dirty Industry, China Plots Course for Green Growth', *Guardian* (5 February: 29)
Watt, N., Laville, S. and Dodd, V. (2011) 'Too Few, too Slow, too Timid: Tories Attack Police over Riots', *Guardian* (12 August: 2)
Wilkinson, R. and Pickett, K. (2009) *The Spirit Level: Why Equality is Better for Everyone* (London: Penguin)
Williams, R. (2010) 'Every Little Helps: Council Plans "Big Society" Reward Points', *Guardian* (1 November: 1)
——(2011) 'Civic Spirits', *Guardian* (9 March: 1)
Willsher, K. (2011) 'Man Refused French Nationality over "Sexist" Views', *Guardian* (10 June: 21)
Wilson, R. (2010) 'Show us the money', *Guardian* (27 October: 3)
Wintour, P. (2010) 'Residents to vote on plans for Mayors', *Guardian* (20 November: 28)
——(2011) 'Purnell Urges Rethink Over Benefit Payments Policy', *Guardian* (30 April: 16)
World Bank, The (2010a) *Cities and Climate Change: An Urgent Agenda* (Washington, DC: World Bank)
——(2010b) *Eco^2 Cities: Ecological Cities as Economic Cities* (Washington, DC: World Bank)
World Clean Energy Awards (2007) 'Award Winners 2007: City of Vaxjo, Sweden', http://www.cleanenergyawards.com/top-navigation/nominees-projects/nominee-detail/project/29/?cHash=fccfe22068 (accessed 19 May 2011)

World Mayors Council on Climate Change (2010) 'Mexico City Pact', http://www.worldmayorscouncil.org/the-mexico-city-pact.html (accessed 22 June 2011)

Zadek, S. and Merme, M. (2003) *Redefining Materiality* (London: AccountAbility/ UK Social Investment Forum)

——and Radovich, S. (2006) *Governing Collaborative Governance: Enhancing Development Outcomes by Improving Partnership Governance and Accountability*, AccountAbility and the Corporate Social Responsibility Initiative, Working Paper no. 23 (Cambridge, MA: John F. Kennedy School of Government, Harvard University)

INDEX

Note: Italicised page numbers refer to illustrations.

African Charter on Human and Peoples' Rights, 52
Aldersgate Group, *19*
American Declaration of Rights and Duties of Man (USA 1948), 52
Amman (Jordan), *Refugees and the pressure of rapid urbanisation*, 67
Amsterdam (The Netherlands) *Food, rural–urban linkages and competitiveness* case interview, 106–8
Anderson, Samantha, 40–41
Arab Spring, 24–25
area-based negotiations, 44–50
Asuncion (Paraguay), *Bringing informal waste collectors into the mainstream*, 90–91
Australia; Citizenship Act (2007), 52; *Preventing violence before it occurs* case interview, 21–23

banks; and community land trusts (CLTs), 92–94; diversity in, 91–95; protests against, *86–87*; and regulatory reform, 87–90
The Basque Country (Spain), *Economic and democratic renewal through enterprise*, 45–46
Benello, C. G., 11–12
biosphere reserve, 46–50
Björkmarker, Åsa Karlsson, 27–28
Bolivia, policy developments in, 57
bottom-up community provision, 37–38
Brighton and Hove (UK) *Towards a social contract to preserve the biosphere* case interview, 46–50
Brugmann, J., 13

California (USA), *Setting the nation's highest renewable power target*, 55–56
Canada, *Creative use of technology*, 108–9
Cape Town (South Africa) *A constructive constitution* case interview, 53–55
Caracas (Venezuela), *Bottom-up community provision*, 37–38
carbon emissions, *74–75*
Cartwright, Anton, 54–57
case interviews; Amsterdam (The Netherlands) *Food, rural-urban linkages and competitiveness*, 106–8; Brighton and Hove (UK) *Towards a social contract to preserve the biosphere*, 46–50; Cape Town (South Africa) *A constructive constitution*, 53–55; Madison (USA) *Community housing*, 92–94; Moratuwa and Matara (Sri Lanka) *Centring women in governance*, 39–41; Toyama (Japan) *The challenges of compact policies*, 68–70; Växjö (Sweden) *Common concerns and action on climate change*, 27–29; Victoria (Australia) *Preventing violence before it occurs*, 21–23
Centering Women in Reconstruction and Governance Project (CWRG), 39–40
Changwon (Republic of Korea), *Declaration of an environmental capital*, 106

children; and poverty, *17*, 17–18; traffic deaths, *66*
China; five-year plan 2011–15, *73*; *No promotion for local officials who do not care for their family*, 46; Paying farmers to move, 65–66
Cisco's Connected Urban Development (CUD), *80*, *81*
cities; compactness, 63–71, *64*, *66*; economic power of, 63–64, *64*; mayors, *119*, 119; and resource flows, *100*, *101–102*
Citizenship Act (Australia 2007), 52
civil liberties, 24–25
climate change, 27–29
Columbia, *Creative use of technology*, 108–9
community groups, and public services, 35–38
community land trusts (CLTs), 92–94
Community Pool Resources (CPRs), 43–44
compactness; challenges, 68–70; need for, 63–71, *64*, *66*
Conaty, Pat, 93
constitutions, harmonised, 51–59
contributions; and fairness, 42–44; partner, *119*, 119–21, *120*
corporate social responsibility (CSR) programmes, analysis of, *79–80*, *81*
Crockett, Thurstan, 47–49
Croydon (UK), *Early interventions in social care*, 109–10
Cuba, *Lesser government and bigger society*, 36–37

debt crisis, eurozone, 17
Delhi (India), *Green does not always mean good*, 70
density, 63–71, *64*, *66*
development, urban, 72–85, *73–77*, *79–81*
domestic violence, 15. *see also* women

economic power of cities, 63–64, *64*
ecosystem services, 82–85
Ecuador, *Ecosystems services*, 82–85
empowerment; embedding and maintaining, 114–17, *115–116*; outcomes from, 38–41
environmental limits; legislation, *57*; meeting, 16–18, *17*
environmental significant behaviour (ESB), 29–31
equality illusion, 15. *see also* women
eurozone debt crisis, 17
excluded groups, mobilising, 45

fairness, and contribution, 42–44
families, care for, 46
financial crisis, 18
Finland, *Creative use of technology*, 108–9
Forum for the Future's Sustainable Cities Index, *79*, *81*
fuel prices, 65
Fund for the Protection of Water (FONAG), 83

gas prices, 65
Germany, *'Carrot and stick' citizenship schemes*, 44–45
global financial crisis, 18
Green Alliance, *91–92*
green concentrate, 103–4, *104*
green economy; approach to, *76*; definition, 78–82, *79–80*, *81*; Pennsylvania (USA), 77–78; swot analysis, 77
GreenWorks, 78

Haldane, Andrew, 91
happiness, concept of, 26–29
Havana (Cuba), *Lesser government and bigger society*, 36–37
Hebei (China), *No promotion for local officials who do not care for their family*, 46
Home Depot Foundation's Sustainable Cities Institute, 18, *79–80*, *81*
Hungary, policy developments in, 56
Huyton (UK), *'Carrot and stick' citizenship schemes*, 44–45

ICLEI–Local Governments for Sustainability, 112
India, *Green does not always mean good*, 70
industrialisation, waves of, 19–21, *20*
informal waste collectors, 90–91
infused resilience, taking action toward, 118–21, *119*, *120*
International Centre for Sustainable Cities (ICSC), 39–40
intervention matrix, 103, *103*
irrigation, 101

Japan, *The challenges of compact policies* case interview, 68–70
Jätkäsarri quarter, Helsinki (Finland), *Creative use of technology*, 108–9
Jordan, *Refugees and the pressure of rapid urbanisation*, 67

Kenya, policy developments in, 56–57
Kunstler, J. H., 12–13

Lagos (Nigeria), *Changing the lives of millions through the railway*, 66
land use, young people's insights on, 45
leadership, local, 118–21, *119*, *120*
Lima (Peru), *Wastewater reuse for irrigation*, 101
literature review, 9–14
local leadership, 118–21, *119*, *120*
local spending, 102–5, *103*, *104*, *105*
Local Strategic Partnership (LSP), 48–49
Localism Act, (UK 2011), 35–36
London (UK), *Mobilising excluded groups in planning and campaigning*, 45
London Citizens, 45
London School of Economics (LSE), 67–68
Longlands, S., 11
Los Angeles (USA), *Separate lives: lessons on riots and gangs*, 30–31

Madison (USA) *Community housing* case interview, 92–94
Manchester (UK), *Local buying to save money and improve communities*, 102–5
Matsumoto, Tadashi, 69
mayors, world, 119, *119*
McInroy, N., 11
McRae, Jane, 40
Medellin (Columbia), *Creative use of technology*, 108–9
Michigan (USA), *Place-based governance*, 110
Middle East, 24–25
migration, incentivised, 63–71, *64*, *66*
Millennium Development Goals (MDGs), report card, 15–18, *17*
Moratuwa and Matara (Sri Lanka) *Centring women in governance* case interview, 39–41

national salary insurance (UK), 42–43, *43*
need, establishing, 9–14
negotiations, area-based, 44–50
Nepal, *Bottom-up community provision*, 37–38
The Netherlands, *Food, rural–urban linkages and competitiveness case interview*, 106–8
Neustaudt an der Weinstrasse (Germany), *'Carrot and stick' citizenship schemes*, 44–45
New Orleans (USA), *Saving residents money as gas prices rise*, 65
Nigeria, *Changing the lives of millions through the railway*, 66
nudge theory, 44

OECD, Better Life Index, 26
Ontaria (Canada), *Creative use of technology*, 108–9
Ordos City (China), *Paying farmers to move*, 65–66

Parades (Portugal), *Creative use of technology*, 108–9
Paraguay, *Bringing informal waste collectors into the mainstream*, 90–91
partner contributions, *119*, 119–21, *120*
Pennsylvania (USA), *Green works for the economy*, 77–78
Peru, *Wastewater reuse for irrigation*, 101
Philips Center for Health and Well-being, *80*, *81*
Pickett, K., 12
Polish constitution, 52
Portugal, *Creative use of technology*, 108–9
poverty, child, *17*, 17–18
power, shifts in, 38–41
privatisation, 35–38
public services, decentralisation of, 35–38
public–public water partnerships, 94
Purena (Nepal), *Bottom-up community provision*, 37–38

Quito (Ecuador), *Ecosystems services*, 82–85

railways, and migration, 66
regulatory reform; and banks, 87–90; demands for, *19*
renewable energy, 55–56
Republic of Korea, *Declaration of an environmental capital*, 106
resilience; interpretation of, 112–14, *113*, *114*, *115*; literature review, 9–14
resource flows, 99–102, *100*, *101–120*
responsibility, devolving, 35–41
revolutions, 24–25

San Carlos (USA), *Mobilising excluded groups in planning and campaigning*, 45
Santa Cruz (Bolivia), *Public–public partnership for water*, 94
Shaw, K., 10–11
Silicon Beach, 46–50
smart cities initiatives, 18
smart density, 63–71, *64*, *66*
social networks, 24–25
South Africa, *A constructive constitution case interview*, 53–55
Southern Cross, near financial collapse of, 72

Spain, *Economic and democratic renewal through enterprise*, 45–46
Sri Lanka, *Centring women in governance case interview*, 39–41
sub-prime mortgage crisis, 88
subsidiarity, journey to, 51–56
Suffolk (UK), *Lesser government and bigger society*, 36–37
sustainability, normalising action on, *105*
Sweden, *Common concerns and action on climate change* case interview, 27–29
Swiss Re, *101–102*

taxpayer bailouts, 86
Texas (USA), *Bottom-up community provision*, 37–38
Totnes (UK), *Transition town*, 58–59
Toyama (Japan) *The challenges of compact policies* case interview, 68–70
traffic deaths, of children, *66*
Transition Town Totnes (TTT) initiative, 58–59
Troy Gardens (Madison USA), 92–94

UNEP's International Resource Panel (IRP), and resource flows, 100
UN-Habitat, 13–14
unifying beliefs, 26–29
United Cities and Local Governments (UCLG, 2007), 53
United Kingdom (UK); approach to green economy, *76*; *'Carrot and stick' citizenship schemes*, 44–45; *Early interventions in social care*, 109–10; *Lesser government and bigger society*, 36–37; *Local buying to save money and improve communities*, 102–5; Localism Act (2011), 35–36; London School of Economics (LSE), 67–68; *Mobilising excluded groups in planning and campaigning*, 45; national salary insurance, 42–43, *43*; protests against banks, *86–87*; Southern Cross, 72; *Towards a social contract to preserve the biosphere case interview*, 46–50; *Transition town*, 58–59
United States (USA); approach to green economy, *76*; *Bottom-up community provision*, 37–38; *Community housing* case interview, 92–94; *Green works for the economy*, 77–78; *Mobilising excluded groups in planning and campaigning*, 45; *Place-based governance*, 110; protests against banks, *86–87*; *Saving residents money as gas prices rise*, 65; *Separate lives: lessons on riots and gangs*, 30–31; *Setting the nation's highest renewable power target*, 55–56; sub-prime mortgage crisis, 88
urban development, 72–85, *73–77*, *79–81*

values, common set of, 24–31
Växjö (Sweden) *Common concerns and action on climate change* case interview, 27–29
Venezuela, *Bottom-up community provision*, 37–38
Vermeulen, Pim, 107–8
Victoria (Australia) *Preventing violence before it occurs* case interview, 21–23
violence towards women, 15. *see also* women

waste collectors, 90–91
wastewater reuse for irrigation, 101
WBCSD Urban Infrastructure Initiative (World Business Council for Sustainable Development), 18
Wijerathne, Sumana, 40
Wilkinson, R., 12
women; care for, 46; *Centring women in governance* case interview, 39–41; domestic violence, 15; involvement of, 39; *Preventing violence before it occurs* case interview, 21–23; status of, 15
World Business Council for Sustainable Development (WBCSD), *79*, *81*
world carbon emissions, *74–75*
world population, distribution of, 63
World Wide Fund for Nature (WWF), 16